BIOCHEMICAL STUDIES ON THYROID CARCINOMA

SAMI AL-MUDHAFFAR
HUDA HAMED HASAN

Introduction

Introduction

1-1   Tumor marker: –

Tumor Markers are substances produced by specific tumors. Benign or malignant and by the host (carrier of the tumor). These markers are of different chemical nature; play various functional roles in several important problems related to the diagnosis, treatment and management of tumor [1]. These markers may also be present in the circulation, body fluids or tissues [2]. Many tumor markers have been investigated in the hope of the finding a suitable tool for early detection of cancer but only a few tumor markers have a special position in clinical oncology [3-5].

There are different techniques which could be used for measuring tumor markers such as immunocytological immunohistological methods i.e., immunofluresence used to investigate those markers present only on cell membrane and cytoplasm [6]. Circulated tumor markers can be measured chemically or immunologically, i.e. , radioimmunoassay flurescence assay . [4]

1-2   Sialic acid (s): –

The term sialic acid represents a group of nine carbon poly hydroxy keto acids called neuraminic acid . Sialic acids are N and O substituted dervatives of the parent neuraminic acid [7-8] and usually located terminally as non reducing residues of the carbohydrate prosthetic group of numerous glycoproteins [9] . It is constituent of mucoproteins, mucolipids and lipoprotein - carbohydrate complex [10]. Figure 1 (A and B). In the light of diversity of sialic acids, their properties have been used to give an importance in many forms of life [11-12]. It is an acidic sugar frequently found as a component of eukaryotic carbohydrate structure and prokaryotic cells . It has also been found as a constituent of capsular polysaccharide of pathogenic bacteria [13] .

The most abundant structure of sialic acid is N-acetylnuraminic acid with two importante positions C_2 , C_5 which are glycosidically linked to galactose , glucose [14] .

In N-acetyl nuraminic acid the carbohydrate chain is linked to Asparagine moiety and the N-acetyl group will be terminal as N-glycans but in O-glycans the linkage of carbohydrate chain is linked to threonine or serine moiety figure (2) .

Chair form

α -Hemiaketal form   Open chair form

Figure (1.A) : Different structure forms of sialic acid .

1-2-1 Physical and chemical properties of sialic acids:-

Sialic acids are colorless, doesn't exhibit mutarotation in aqueous solution [15], (Fig 1A). Sialic acids are water soluble, unstable in both acid or alkaline media due to the presence of carboxyl on C_2 (pKa value of 2.6 regarded strong acid), so aqueous solutions decompose when left for some time at room temperature. A great tendency of sialic acid appears to form esters during crystallization from water free methanol[16].

1-2-2 Sialic acids functions: -

The universal component of cell membrane is sialic acid [17-19], (70% of cell surface glycoprotein [20]) and the negative charge on some cells (RBC , platelets) are mainly due to sialic acid [21-22]. It's usefulness involved a cellular contact, cell to cell interaction forming a basic of RBC & platelets agglutinate [19,23]. Sialic acid of RBC surface is important in recogntion , the removal of sialic acids from its surface enhances the binding of influenza viruses to the RBC [24-25]. Sialic acids are useful to maintain in the shape of the normal erythrocytes [26-27]. The removal of sialic acid destroys the ability of TSH to reach to the target cells (follicular cells), so sialic acid seem to posses TSH receptor recognition function [28-29].

Sialic acids act as protective of the plasma glycoproteins from splitting by proteolytic enzymes [30-32], ie, in mucin, sialic acids known to give the mucin it's viscosity and protect it from proteolytic splitting [30-34]. Different types of cancers have been shown in structural alteration at the cellular membranes. These membranes contain sialic acid in the forms of glycoprotein . [35,36]

Accordingly the content of sialic acids have stimulated the measurment of their level of sialoglycoprotein and sialoglycolipid in cancer cells as tumor markers . The recent studies were concentrated on monitoring the total or bound sialic acid in cases of breast cancer [41,207], malignant melanomas [42], kidny skin, liver [43-44], bone [45], ovary carcinoma [46], cardiovascular mortality [132] and other types of cancer when compared to normal persons [206]. Elevated results have been reported to be abnormal in such human cancer [47-49].

Poly sialic acid level was considered as a useful test in various types of malignant thyroid tumors which was measured by immunohistochemical techniques to differentiate different thyroid carcinoma [40]. The elevation of lipid associated sialic acid is used as marker of leukemia, gastrointestinal, of cancer patients [50].

Serum lipid associated sialic acid (LASA) are useful monitor of tumor burden [39-51], and it is propotional to the tumor stage [40-52]. LASA elevation are relaively non specific with respect to type of cancer for routine cancer detection [41-53].

The elevation of lipid associated sialic acid is used as marker of leukemia , gastrointestinal , of cancer patients [38] .

1-2-B   Distrbution of sialic acid in tissues and body fluids:-

Sialic acid occur widely in mammalian principally in mucoids , mucins and glycoprotein in association with galactose , mannose [54-55]. It has been found that N-acetyl neuraminic acid is distrbuted in brain tissues in the form of protein bound and lipid bound sialic acid [56-57]. it is also found in other tissues such as skin kidney [43] thyroid [58] breast colon , liver [44] , lung , ovary and stomch .

The sialic acid content of serum is distributed between serum protein functions (α1,α2,β,λ globulins). α2- globulin fraction of serum as a highest content of sialic acid and that 90% of sialic acid found in a combined α and β globulin.

Larent [61] and Montreil [62] have studied qualitativly and quantitatively of serum carbohydrate complex . Schults [163] observed elevation in cancer disease regarding the elevation of carbohydrate bond to the α2- globulin fraction in serum.

This may reflect a general pattern of change in tumor glycoprotein composition linked to increase malignant expression [64].

Human urine content of sialic acid is too low to be estimated. Cerebrospinal fluid content of sialic acid under pathological conditions resulted from the disturbance of protein bound sialic acid and seromucoid determination in CSF is of limited clinical value [65,204].

Sialomucin manifest on some types of cancer as surface consistuent [66] by masking tumor and histocompatibility antigens, this apper to provide protective mechanism for cancer cell [68].

1-3   Glycoproteins :-

Glycoproteins are conjugated proteins containing prosthetic groups of one or more of heterosaccharide(s), the later is usually branched, lacking repeating units, and bound covalently to the peptide chain [69-73] Yet the branched chain linked to protein is composed of galactose or mannose, glucos amine or galactosamine, fucose (which   is 6-deoxygalactose) and sialic acid [72] . The terminal of oligosaccharide chains, is usually connected with subterminal galactose , (Gal.) or N-acetylgalactosamine (Gal Nac), while other sugars present internally [60,75,76] (Figure 3).

1-3-1 Chemical characteristics :-

Glycoproteins have a great variation in chemical or physical properties, according to their location and function [77-78].

Although it is possible to dissociate the polysaccharides chain , complex by pH change or ionic strength yet it is not possible to separate carbohydrate from peptide portion of glycoproteins without directly degrading the entire molecule , therefore carbohydrate are integral part of the glycoprotein [79] .

1-3-2 Glycoprotein functions :-

Glycoproteins serve as transporter for vitamins, lipids, minerals, trace elements, and hormones, ie prealbumin responsible for the transport of thyroxine and vitamin A [80]. Glycoprotein act as intermediary information processers (receptors) for hormone molecules [81-82] . Progesterone receptors (important for the managment of certain types of cancer) [80,83]. Also they act as enzymes; such as nucleases, hyd.. rogenases enzymes [83] Glycoproteins serves as cellular interaction [84], i.e. cell - cell interaction to form tissue (collagen), and as immunological materials (antigens). [85-87]

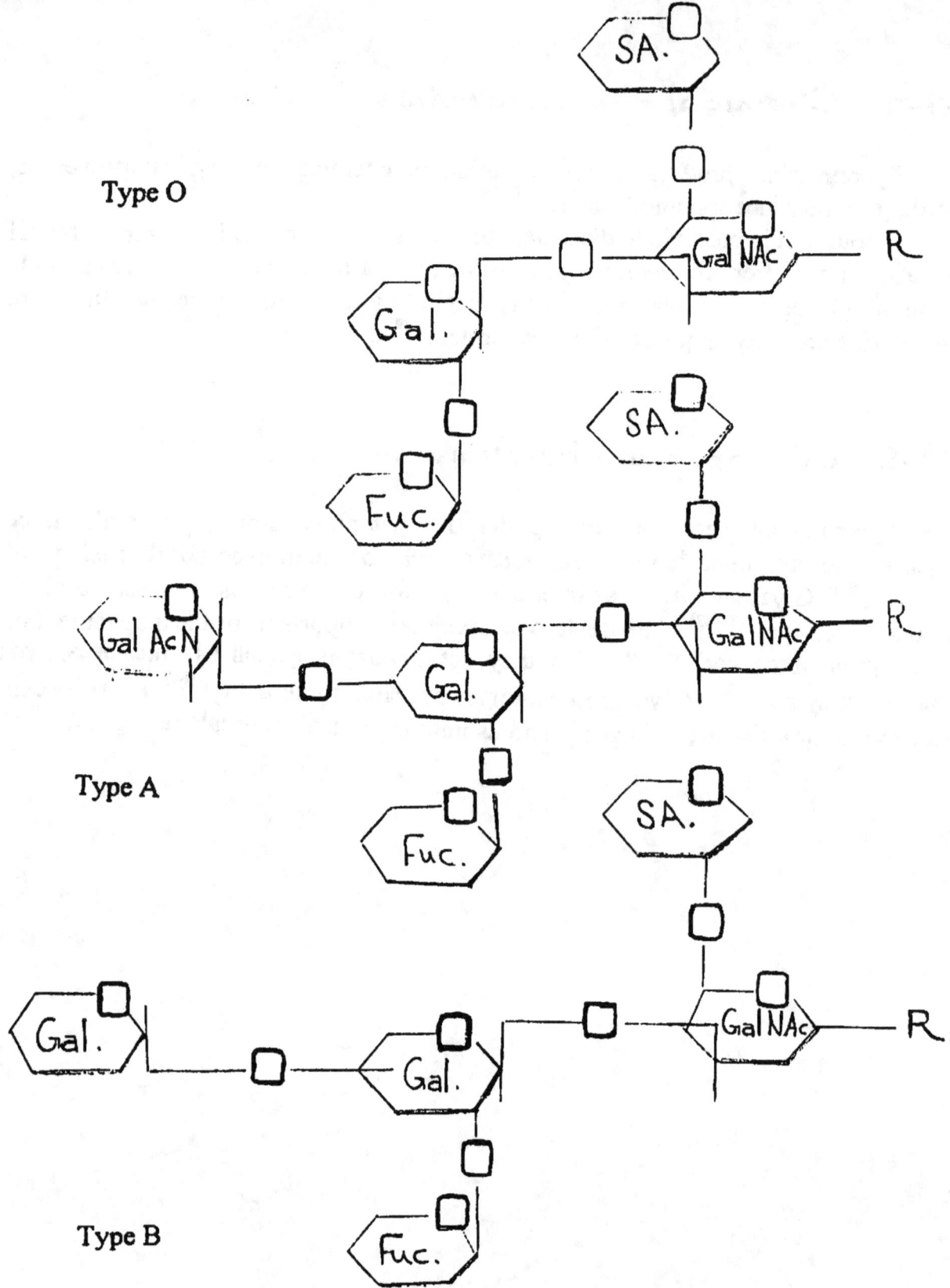

Type O

Type A

Type B

Figure (3) : Blood group antigen as glycoprotein structure .

where SA = sialic acid   Gal NAc= Galactose N- acetyl

Gal = Galactose   Frc = Fructose

1-4 Lectins :-

Lectins are classes of proteins, that interacts with glycoproteins and glycolipids[88] of multivalent carbohydrate binding with at least two sugars, they may be soluble or membranes, with the ability to agglutinate erythrocytes of certain blood group [89-90]. Besides they are specific receptor proteins present in plants (usually their seeds), in vertebrate, animal (snail) and in lower vertebrate (fish ova) [91,5,207].

These properties enable the lectins to discriminate the animal and plant cells or animal cell and tumor cells [92]. The high specificity of carbohydrate binding lectins makes this binding interaction an excellent model for investigation the effect of various ligands on the binding through thermodynamic studies of ligand to protein [93]. These carbohydrate binding specficity of lectin are usefull tool in biochemical and seriological analysis studies by hemagglutinatioin reaction with blood group [94], lectin hemagglutination show a marked differences in agglutinability between normal and malignant cells [95]. So lectins is considered as a discriminating markers for neoplastic cells and tissues [94,96].

There are many types of lectin binding with follicular adenoma, Wheat germ agglutinine, Concanavalin A agglutinine, Ulex europaeus agglutinine I, Peanut agglutinine, Ricinus communis agglutinine and Helix pomatia agglutinine [70-71].

1-4-1 The roles of lectins: -

Lectins have two critical properties, specificity for sugar residue, and bivalancy or polyvalency [72]. It seems to be involved in inter-action between cells and proteins of intracellular matrix, such as collagen, and help to maintain tissue and organ structure [73]. In normal thyroid glands, Ulex europaeus agglutinine I (Lectin) is able to discriminate between follicular cells and C-cells, when benign and malignant tumors are compared for Ulex europaeus agglutinine I affinity [70]. There is a significant greater frequency of malignant tumor with UEAI - receptor [74].

Recently Lectins act as immunoglobulin - independent defense molecules due to a complement - mediated mechanism . However, non specific immune mechanisms in which Lectins play a role have been shown to be important in the neutralization and elimination of various micro organisms in young children,whom does not have material antibody or have not developed efficient systems for specific protection [75].

Liver Lectins seem to be specifically to the face of the hepatocyte with indirect contact with the blood flow. This specific localization would argue for the primary role of partially degraded glycoproteins or possibly the role in the intracellular transport [76]. Lectin properties make them useful for many purposes including blood group typing [77].

1-4-2   Sialic acid binding Lectin :-

Sialic acid binding lectin appear to be mainly presented in vertebrate and dose not present in plants because plants do not contain sialic acid.

Sialic acid binding lectin have been purified and characterized [43] from cancerous human kidney tissue , the purification was by affinity chromotograghy .

There are specific lectins which show a broader range of specificty for sialic acid and its derivatives such as 4 and 9 - 0 - acetylated forms .

The 0 - substituted sialic acid (0 - Ac - NeAc) exhibit intersting species and tissue specific distribution [78],[105] . In mammals , sialic acids of some gangliosides have been reported to be 0 - acetylated such as 9 - 0 - Ac - Neu Ac from malignant myeloma cells [79],[106]. The degree of O-Acetylation changes with transformation or other alteration in the environment of the cell [107]. Other studies indicate that sialic acid binding Lectin (Achatinin) from the hemolymph of Achatin fulica snail, is found to be high specific for 9-O-acetyl sialic acid . This agglutinin agglutinate only those erythrocytes which are known to contain 9-O-Ac Neu Ac [108].

1-4-3   Lectin binding glycoproteins :-

The specificity of binding sites of lectins suggests that there are endogenous saccharides receptor in the tissues from which they are derived, or on other cells or glucoconjugates with which the lectin is specialized to interact . Lectin on the surface of hepatocytes binds terminal galactose residues of a variety of sialoglycoproteins [102]. The fact that lectins have more than one carbohydrate binding sites suggests that they act as to cross - link glycoprotein and glycolipid in the membrane of the same cell for various organization purposes [109] . Binding of glycoproteins to hepatocytes is absolutely depenent on Ca^{++} ions and is completely inhibited by treatment of hepatocytes or the Lectin with neuramidase [17,74].

The removal of sialic acid from the glycan-chains (carbohydrate) of the Lectin induces self - aggregation through interaction with exposed galactose terminal residue [102,110].

1-4-4   Lectin applications :

There are several ways to use lectins for the purpose of probing of the cell surface exposed for analysis and purification of glycoprotein for the study of membranes of a large variety of animal cells, microorganisms and plants [111-112]. Lectin have been used to identify mono and oligosaccharid on cell surface glycoproteins, followed directly from the original red cell agglutination. Furthermore, the saccharide - binding properites of lectins may advantageously be utilized in a number of special analytical and preparative techniques for purification, characterization and sequencing of carbohydrates, glycoproteins and glycolipids [93].

It has been reported that the process of differentiation in embryonic and adult cells accompanied by changes in distribution and number of surface lectin binding sites [113],[114].

And finally lectins are useful as drug carriers like conjugates of concanvalin A with anti-tumor drug such as methotrexate[115].

1-5   Agglutination

It is a serologic reaction involving clumping of a cell suspension by specific antibody (Lectin), the agglutinine observed when particular antigens (such as blood cells or bacteria) are exposed to specific antibody under appropriate conditions by covalent linkage [19,75]. There are many ways to determine agglutination such as those of direct and indirect agglutination, antiglobulin test and immunohistochemical methods for fluorescent antibody [86].

1-5-1   Antigen-antibody agglutination reaction :-

It is generally used to detect antibody to interinsic antigens on RBC surface [87-88]. A major potential application of the agglutinine it is use (coupled to histological markeres) for localizing the presence of bound NANA in tissues [89]. Substances in tissues which react with agglutinine might be distinguished by digestion with neuraminidase or glucoronic acid; which have been used successfuly to remove sialic acid residue that, which react with this agglutinine [90-93].

1-5-2 The agglutination process in tranfusion reaction:-

The mismatching of blood by mixing anti-A or Anti-B with RBC containing A or B agglutinogens respectively, usually occur when the agglutinines attach themselves to RBC, which are (Ig G & Ig M types) thereby the cells clumping to each other, these clumps plug small blood vessels through the circulation [94]. Agglutinine is very important in different blood groups which are immunogenic determinants [95], all blood group have sialic acid terminal bound to lectin [17-63], and they are different in type and arrangment of other terminal sugars present with the side chain of glycoprotein such variation leads to the presence of different blood group and it's specificity of the individuals [96-98]

1-6 Thyroid gland:-

Thyroid gland is composed of two lobes, one on each side of the neck, located in the anterior triangle of the neck in front of larynx and trachea Fig. (4) joind by the thyroid isthmus [128-129]. These lobules comprised of about 20 - 40 follicles which are lined by epithelial cells that surround deposits of colloid, and produce thyroid hormones. Calcium oxalate crystals are commonly seen within the colloid in (40-79)% of the normal glands.[130,131]

Thyroid gland secrete L-T$_3$ and L-T$_4$ into blood [132], where 99.9% of these hormones are bound to protein in blood [131]. The releasing of L-T$_3$, L-T$_4$ in to blood starts with the degradation of thyroglobulin molecules by proteolytic enzymes, this process is stimulated by thyroid stimulating hormone (TSH) of the anterior pituitary gland which play as a key of the physiological activity of the follicular cells [134-135]. Thyroid stimulating hormone secretion is affected by many physiological effect such as iodin deficiency, Iodine is essential for thyroid synthesis 100-150 μg. arc required daily to form normal quantities of thyroid hormones) or partial thyrodectomy [127,136], or thyroid stimulating hormone TSH level (its a glycoprotein of molecular weight approximately 30000 D. consist of two subunit designated α and β [137]). Increasing level of thyroid stimulating hormone stimulating the thyroid to become more active [138].

1-6-1 Thyroid hormone receptors :-

Receptors are exhibited affinity , specificity , saturability and subsequently biochemical events following the interaction with the hormones [139]. Thyroid hormones L-T_3 and L-T_4 bind receptors, either at cell membrane (pituitary hormone receptor) or nucleus the receptor present as free or bind to carbohydrate residue or companied with DNA in the nucleus of the cell [140-142] . Thyroid hormone receptor are important in the clinical diagnosis [143,144] and treatment therapy ; thyroid tumor which do not contain significant amounts of receptors are unlikely to response to endocrine therapy [145].

1-6-2 $T3$, $T4$ and TSH levels in thyroid diseases:-

T_3 contributes significantly to the maintenance of the euthyroid state , and the total T_3 levels has a role in screening for thyroid disease in conjunction with other tests [146]. T_3 alone cann't diagnose hypothyroidism, but it may be more sensitive than thyroxine T_4 for hypothyroidism. In most patients the total T_4 level is a good indicator of thyroid status , however it can sometimes be inadequate , and diagnostic efficiency may be improved by other tests. Serum TSH level are raised in cases of primary hypothyroidism. The diagnosis of hypothyroidism by the finding of a low total or free T_4 value is readily confirmed by a raised TSH level. In hyperthyroidism levels of T_4 and T_3 are elevated and TSH secretion is suppressed [112,134].

1-6-2-1 Hyperthyroidism :-

Thyroid hormones secretion is increased by a number of pathological conditions such as toxic multinodular goiter [112,208], and appears in patients with nodular goiter. Other disorders of this type are toxic adenomas or benign tumors [128] Fig. (5).

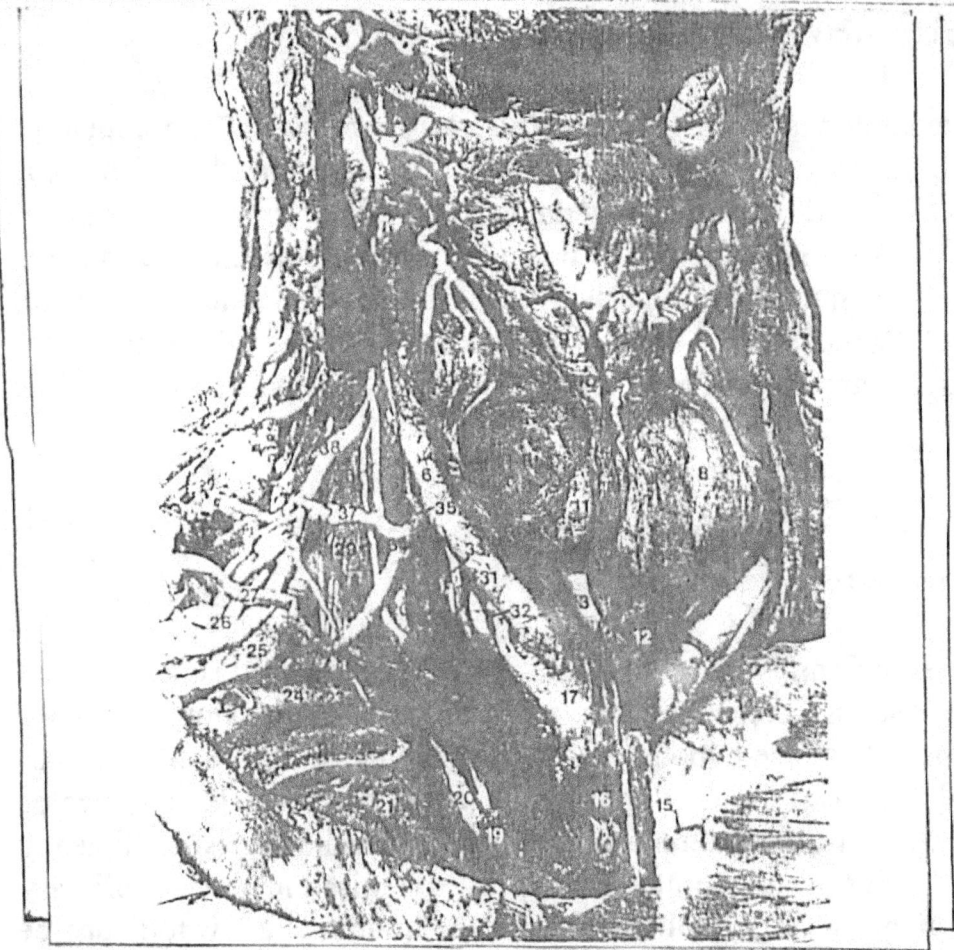

Figure (4) :The thyroid gland.

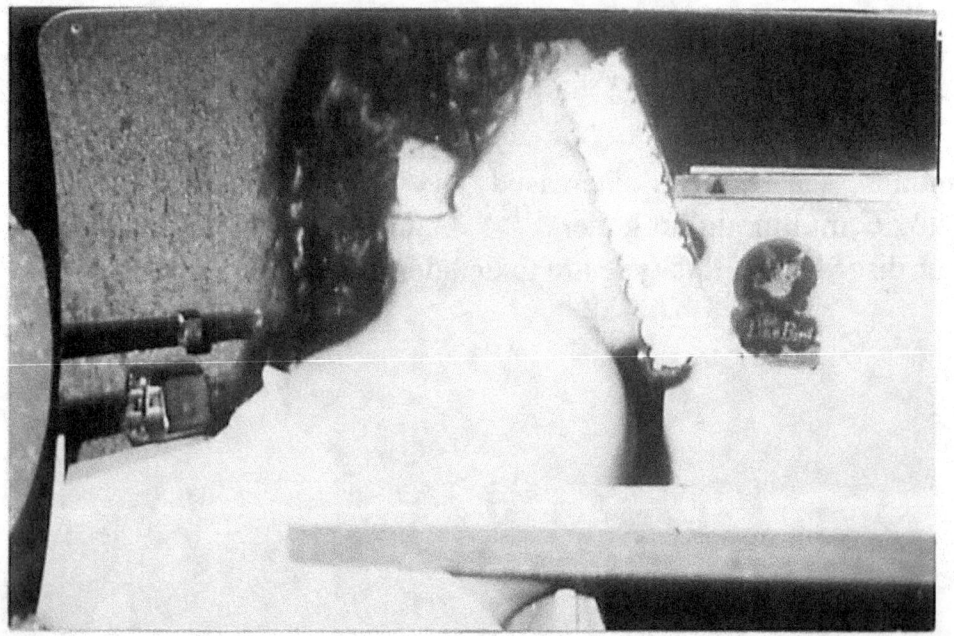

Figure (5) :Hyperthroidism of (43 years female) affected by multinodular goiter

1-6-2-2   Hypothyroidism :-

Thyroid hormones secretion is decreased whenever insufficient amounts of these hormones are available to tissues [144], this will cause tissue damage and reflect clinicaly by enlarged thyroid gland .Hypothyroidism can result from taking therapy, iodine deficiency, or surgical thyroidecomy or other abnormalities (ie, overproduction of TSH) or tumor .

1-6-2-3 Endemic goiter:

Endemic goiter is enlargment of the thyroid gland (when glands appear to weight more than 35 gm readily plapable).

It is secretion of L-T_3, L-T_4 maintains euthyroidism (or non toxic) . It can be localized into non toxic and diffuse types according to the American thyroid association [148-149].

1-6-2-4 Carcinoma of the thyroid gland:

Cancer of the thyroid gland accounts for approximately 0.5% of all cancer deaths. It is more frequent in the endemic goiter in many areas of the world, but this cause is gradually being eradicated by dietary iodine. Thyroid cancer is twice as common in females as in males. The incidence peak is between 40 - 60 years of age. There are many types of this disease ie papillary carcinoma, follicular carcinoma, undifferentiated carcinoma [134].

The induction of carcinoma of thyroid gland by chemicals carcinogen (ie 2-acetylaminofluorene) produce irreversible effect on the epithelial cells lining the follicles, or by iodine - deficient diet , antithyroid drugs , radiation exposures of 350 r carry 30 percent risk of developing thyroid nodules [150]. In general increasing age , cause diffuse goiter to be nodular or multinodular , located deep within gland . The solitary nodule refers to enlargment of thyroid representing benign and malignant tumors , decreased by external radiation therapy for head , neck and chest . Sialic acid is produced by thyroid gland tumor cells its levels may be dependent on the mass of tumor and type [141].

1-7 Radioimmunoassy (RIA) :-

(RIA) method depends on the competition between the thyroid hormones and $[I^{125}]$ labelled hormones for a limited number of binding sites are a specific antibody for each of these hormones. The proportion of $[I^{125}]$ hormone bound to the antibody is inversely realted to the concentration of hormone present in the serum . By measuring the proportion of $[I^{125}]$ hormone bound in the presence of standard sera containing various known amounts of the hormone . The concentration of each of these hormones (T_3, T_4 & TSH) for example in diseased samples can be interpolated .

Materials and Methods

Materials and Methods

2-1 Instruments: -

The instruments used during this study were; - Refrigerator (-20° C), cooling MSE centrifuge, Gamma counter LKB, Biochrom ultrospec 4050, LKB spectrophotometer, Pye - unicam model 294-MKZ pH meter, Memert water bath, SM - Shaker, LKB - Cold room for gel filteration, LKB 2117 Multiphore system electrophoresis system and General Electric scanning.

2-2 Chemicals :-

All common laboratory chemicals and reagents used were of analar grade, unless otherwise specified. Kits of T_3, T_4, TSH and Technetnetium-99m sterile generator were purchased from Amersham company. The concentration of tracer T3, T4 and TSH in each kit were approximately 0 - 12 nmol T3 / L, 0-320 nmol T4 / L and 0 - 50 µIU TSH / ml. respectively.

The following chemicals were obtained from BDH company, UK, glacial acetic acid, $CaCl_2$, methanol, 2-mercaptoethanol, butyl acetate, acetic acid, perchloric acid and Tris (hydroxy methyl - ethyl amine). The following chemicals were obtained from Fluka company, Switzerland, $CuSO_4.5H_2O$, $Na_2HPO_4.7H_2O$, KH_2PO_4, bovine serum albumin, and human serum. Hopkings and Williams company, supply D-fructose, D-mannose, D-galactose, D-Sucrose, D-glucose. From Riedel Deheform company D-Xylose, phosphotungstic acid and glucouronic acid were obtained.

2-3 Chemical preparation :-

Regarding these solvents, were prepared sustained deionized water for dissolving appropriate amounts.

Materials of protein estimation :-

1- Complex - forming reagent : prepared immediately before use by mixing solution A , B and C in proportion 100:1:1 , respectively.

Solution A: 2% (w/v) Na_2CO_3 in D.W. .

Solution B: 1% (w/v) $CuSO_4.5H_2O$ in D.W. .

Solution C: 2% (w/v) Sodium potassium tartarate in D.W. .

2- 2N NaOH

3- Folin reagent use at 1N concentration .

4- Standard bovine serum albumin 4 mg/ml in D.W. .

Materials of TSA and LASA determination :-

1- Chloroform / methanol (2:1 v/v) solution by mixing two volumes of chloroform to one volume of methanol .

2- Butyl acetate / methanol (85:15 v/v) solution by mixing (85 ml) of butyl acetate then adding (15 ml) of methanol .

3- Phosphotungstic acid solution (1 g/ml) :-

Wet 10 gm . then dissolve in 10 ml deionized water , heat gently to obtain clear solution (keep in Refrigerator)

4- $CuSO_4$ (0.1M) : Dissolving 0.04 gm of the salt in 2.5 ml of deionized water .

5- Resorcinal regent :- Ten ml of 2 % (w/v) stock resorcinol + 9.75 ml. deionized water + 0.25 ml of 0.1M $CuSO_4$ brought to final volume of 100 ml. with concentrated HC l prepared daily .

Chemicals preparation of seromucoid and protein bound hexose determination :-

1- NaCl (85%)

2- Perchloric acid (1.8M). Prepared by dilution of 16.6 ml of 64% perchloric acid to 100 ml. distilled water .

3- Phosphotungstic acid (5%) in 2N HCl .

4- Ethanol(95%)

5- NaOH (0.1N) .

6- Stock standard sugars ,100 mg. galactose + 100 mg mannose were dissolved in 100 ml. water saturated with benzoic acid and kept in refrigerator for storage. Working standard : 1 ml. of stock solution + 9 ml. of distilled water and was prepared freshly for use .

7- Orcinol reagent (2%) recrystallized orcinol is dissolved in 30% v/v H_2SO_4.

8- H_2SO_4 (60%).

Buffers preparation of human thyroid gland tumor homogenate:-

The buffer solutions were prepared by dissolving suitable amounts of salts in distilled water and the appropriate pH was adjusted by pH meter .

1- 4 mM β- mercaptoethanol
2- 20 mM $CaCl_2$: 0.212 gm $CaCl_2$ in 100ml of distilled water .
3- 2 mM EDTA : 0.075 gm in 100 ml of D.W. .
4- 0.02 mM HC l
5- 75 mM KH_2PO_4 : 13.35 gm in 1000 ml distilled water
6- 0.15M NaCl : 8.765 gm in 100 ml of distilled water .
7- 0.9% NaCl .
8- 75mM Na_2HPO_4 : 8.98 gm in 1000 ml of distilled water .
9- 0.02M Tris (hydroxy methyl) amino ethane : 2.432 gm in 1000 ml of D.W.

2-4 Patients & Blood samples :-

Three groups of thyroid patients; hyperthyroidism , hypothyroidism, thyroid carcinoma and normal healthy persons were included in this study, Group I contained twenty patients, Group 2 consisted of sixteen patients and group 3 comprised of eleven patients with cancerous thyroid tumors. In addition group of healthy subject were also included.
All patients were newly diagnosed by thyroidal scanning and biochemical tests . not underwent any type of therapy , patients may interfere with out study excluded . The host information of all patients and healthy subjects is summarized in table (2-1).

2-4-1 Blood samples :-

Whole blood was drawn from the patients in section 2-4 and allowed to clot at room temperature and was centrifuged at 3000 r.p.m. for 10 minutes . The resulting sera were separated and kept at (-20 C) until assayed for biochemical investigations.

Table (2-1) : The host information of thyroid tumor patients and healthy subjects studied .

Patients	Number	Range (year)	Average age(year)	Type of disease tumors
Normal controls	20	(14-50)	32	
Hyperthyroidism	15	(19-65)	42	Nodular goiter
	15	(13-66)	39.5	Multinodular goiter
	20	(4-70)	37	Toxic MNG
	10	(12-50)	31	Diffused goiter
Hypothyroidism	(16)	(13-60)	36.5	
Thyroid carcinoma	11	(28-70)	49	(Pappilation and cold nodular goiter)
Pathological conditions (different diseases with thyroid tumor)	10	(4-60)	32	

2-5 Determination of T3 , T4 and TSH in sera of thyroid diseased patients and healthy subjects by RIA :-

2-5-1 Determination of L-T3 :-

Procedure :-

1- Assemble and label assay tubes .

2- Pipette 50µl of zero standard into NSB (non - specific binding) tubes.

3- Pipette 50µl standard , control or sample into appropriate tubes .

4- Dispense 500µl tracer into all tubes , set aside total count tubes .

5- Dispense 500µl of antibody suspension into all except NSB tubes;
   dispense 500µl NSB reagent into NSB tubes .

6- Vortex, cover and incubate at 37°C for 60 minutes .

7- Attach the rack to the separator base, leave for 15 minutes. Decant
   and drain for 5 minutes with blotting .

8- Place the tubes in gamma counter and count for 1 minute .

Plotting :-

   Standard curve using an RIA curve fit program . Results was calculated using Linear plotting counts against concentration . (Fig 2-1)

2-5-2   L-T4 determination :

Procedure :-
1- Assemble and label assay tubes .
2- Pipette 50μl zero standard into NSB (non-specific binding) tubes .
3- Pipette 50μl standard control or sample into appropriate tubes .
4- Dispense 500μl tracer into all tubes , set aside total count tubes .
5- Dispense 500μl of antibody suspension into all except NSB tubes ; dispense 500μl NSB reagent into NSB tubes .
6- Vortex , cover and incubate at 18-28°C for 45 minutes .
7- Attach the rack to the separator base, leave for 15 minutes , Decant and drain for 5 minutes with blotting .
8- Place the tubes in gamma counter and count for one minute .
9- Plot the data similar to that of mentioned in section 2-5-1 for L-T$_3$ determination Fig. (2-2).

2-5-3   TSH determination :

Procedure :-
1- Label coated tubes for each standard , control and patient sample to be assayed .
2- Pipette 100μl of standard control or patient sample into the appropriate tube .
3- Dispense 100μl of Tracer Reagent to each tube .
4- Vortex gently for 1 - 2 seconds
5- Incubate for 90 minutes at 37°C .
6- Decant or aspirate contents of tubes .
7- Add 2.5 ml wash solution to all tubes and repeat step 6 . Add the wash solution gently to avoid foaming .
8- Repeat step 7 .
9- Repeat step 7. making a total of 3 washes
10- Place tubes in gamma counter and count for one minute .

Plotting :-
     Plot the I^{125} counts (CPM) against the TSH concentration on linear - Linear graph paper Fig. (2-3).

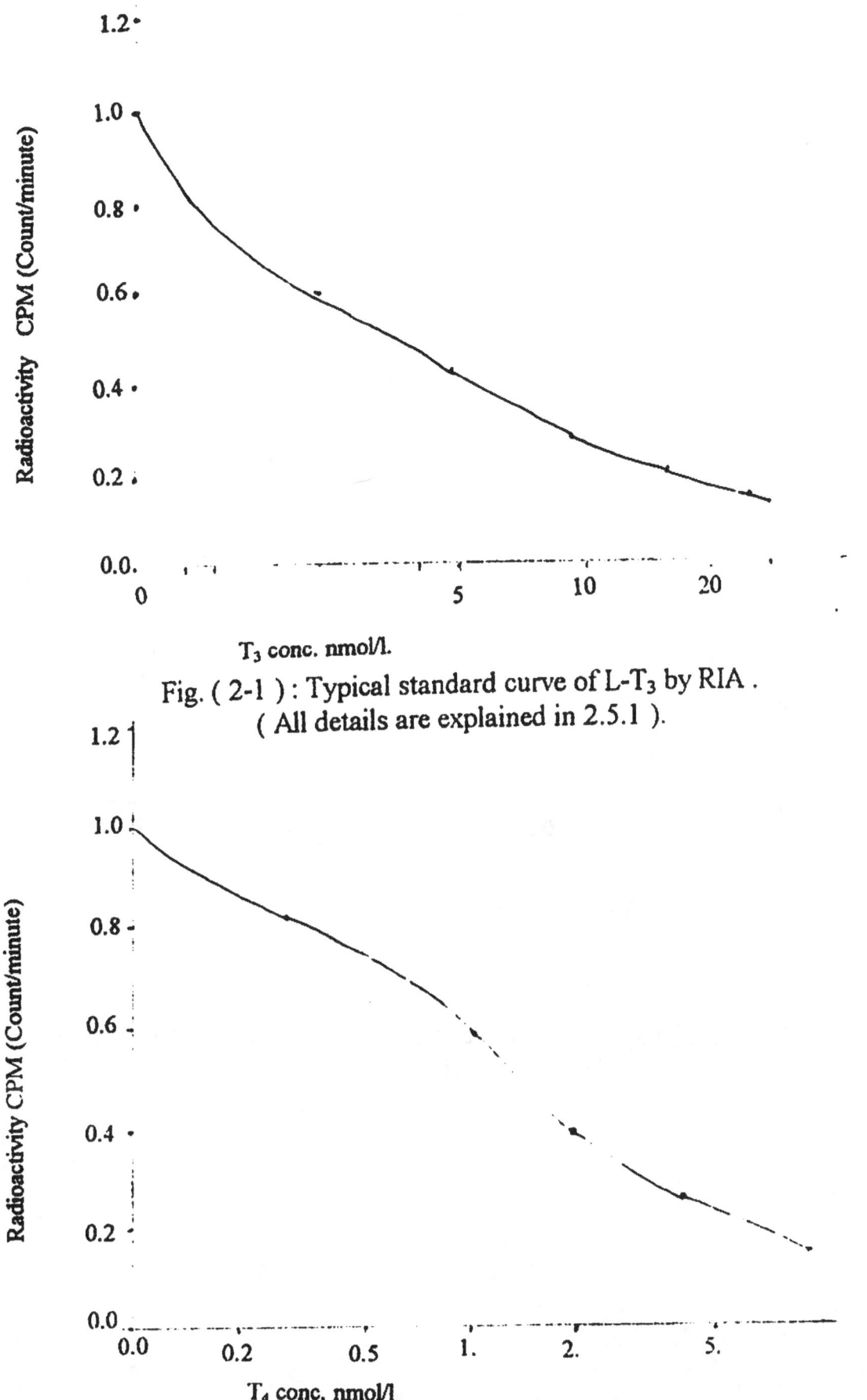

Fig. (2-1) : Typical standard curve of L-T₃ by RIA .
(All details are explained in 2.5.1).

Fig. (2-2) : Typical standard curve of T₄ by RIA (All details are
explained in 2.5.2).

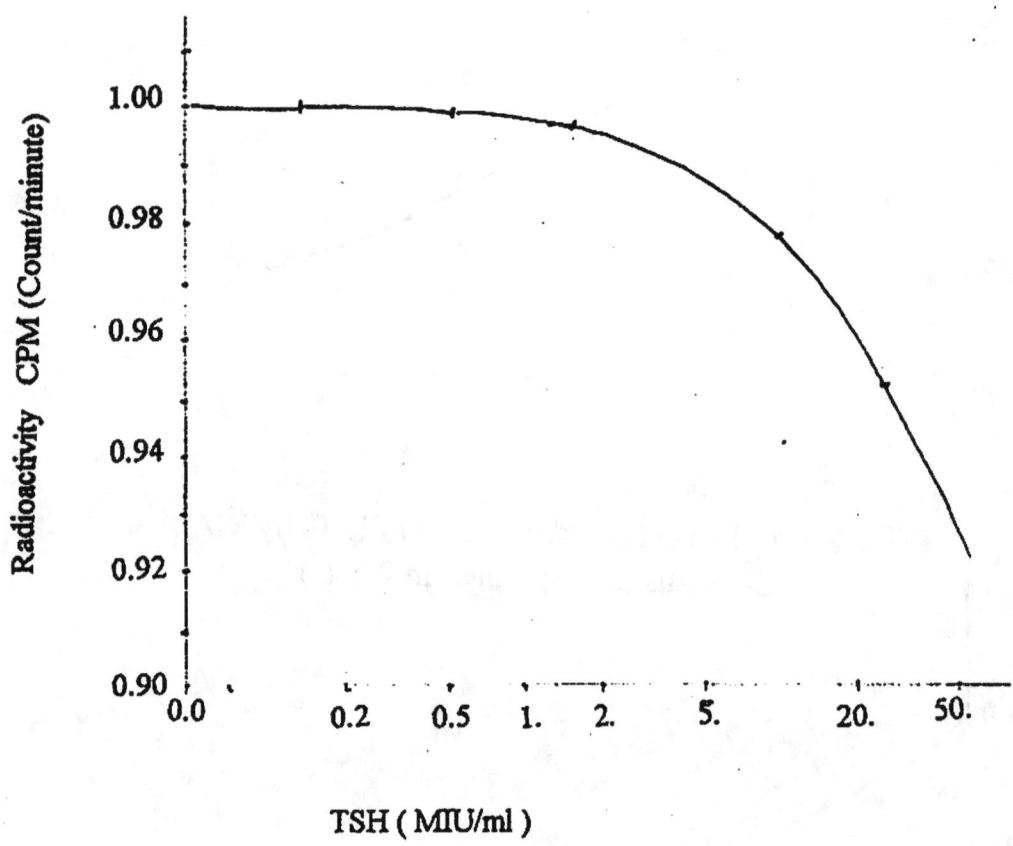

Fig. (2-3) : Typical standard curve of TSH by RIA.(All details are
Explained in 2.5.3).

2-5-4 Thyroid scanning :-

The scan was carried out by Technetium - 99m sterile Generator Amertec 11 taken up by injection 2 milicur. to study the functioning of thyroid tissue. Scintillation scanners moving across the neck and recording the scanning imag instrumentally on paper or film, by maxi camera and formatter. Several projections were used to (view) the neck from different directions.

Estimation of different constituents :-

2-6 Protein estimation :-

Serum total protein were estimated by Lowry *et. al.* [151] . method using bovine serum albumin as standard protein with different serial dilution expressed as (µg/ml) of each sample . The results of the absorbance at 600 nm of standard were ploted vs. standard concentration fig (2-4) .

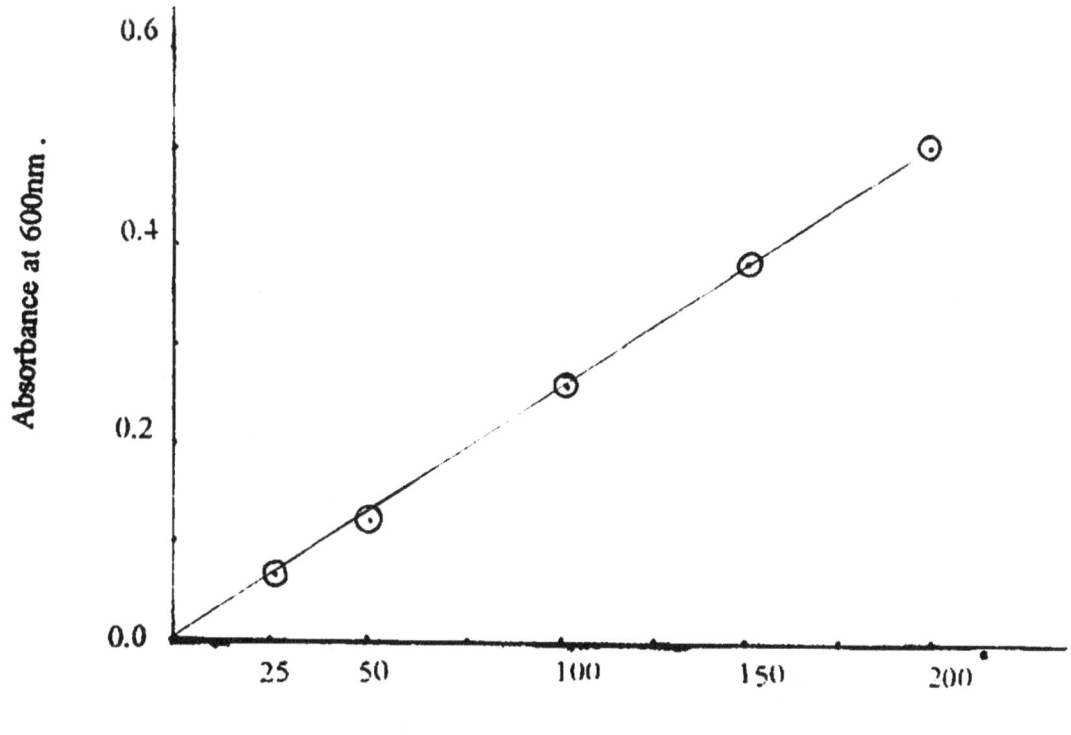

Concentration of (BSA) , (µg./ml.) .

Fig. (2-4) : Standard curve for estimation of total serum protein by (Lowry method) , using (BSA) as standard .

2-7 TSA and LASA determination :-

Katapoids *et. al.* [152] method was used as a procedure for determining LASA and TSA but with slight modification of the volume of sample used . Fifty µl of serum was placed in a test tube having 50µl of water , then vortexed for 5 sec. and placed on ice. Three ml. of chloroform / methanol (2:1v/v) were added at 4°C, then again vortexed and 0.5 ml. of cold water was added, then also vortexed and centrifuged for 5 minutes at 3000 r.p.m. at room temperature. Then one ml. of resulting upper layer was transferred to another test tube, then 50µl of phosphotungstic acid 1 g/ml. was added, then vortexed and allowed to stand for 5 minutes, then these tubes were centrifuged for 5 minutes at 3000 r.p.m. The upper layer was removed but the remaining precipitate was dissolved in 1ml. of deionized water . TSA concentration was determined by taking 20µl of serum and 980µl of deionized water were placed in a test tube, vortexed and placed on ice . To each assay TSA and LASA , 1 ml of resorciol reagent was added for exactly 15 minutes followed by 10 minutes on ice bath, 2 ml. of butyl acetate / methanol (85:15 v/v) was added to each tube , then vortexed and centrifuged for 10 minutes at 3000 r.p.m. The extracted chromphore was read at 580 nm .

Standard curve is prepared by treating standard sialic acid tube as explained above by plotting absorbance vs. N-acetylneuraminic acid concentration , Fig (2-5) .

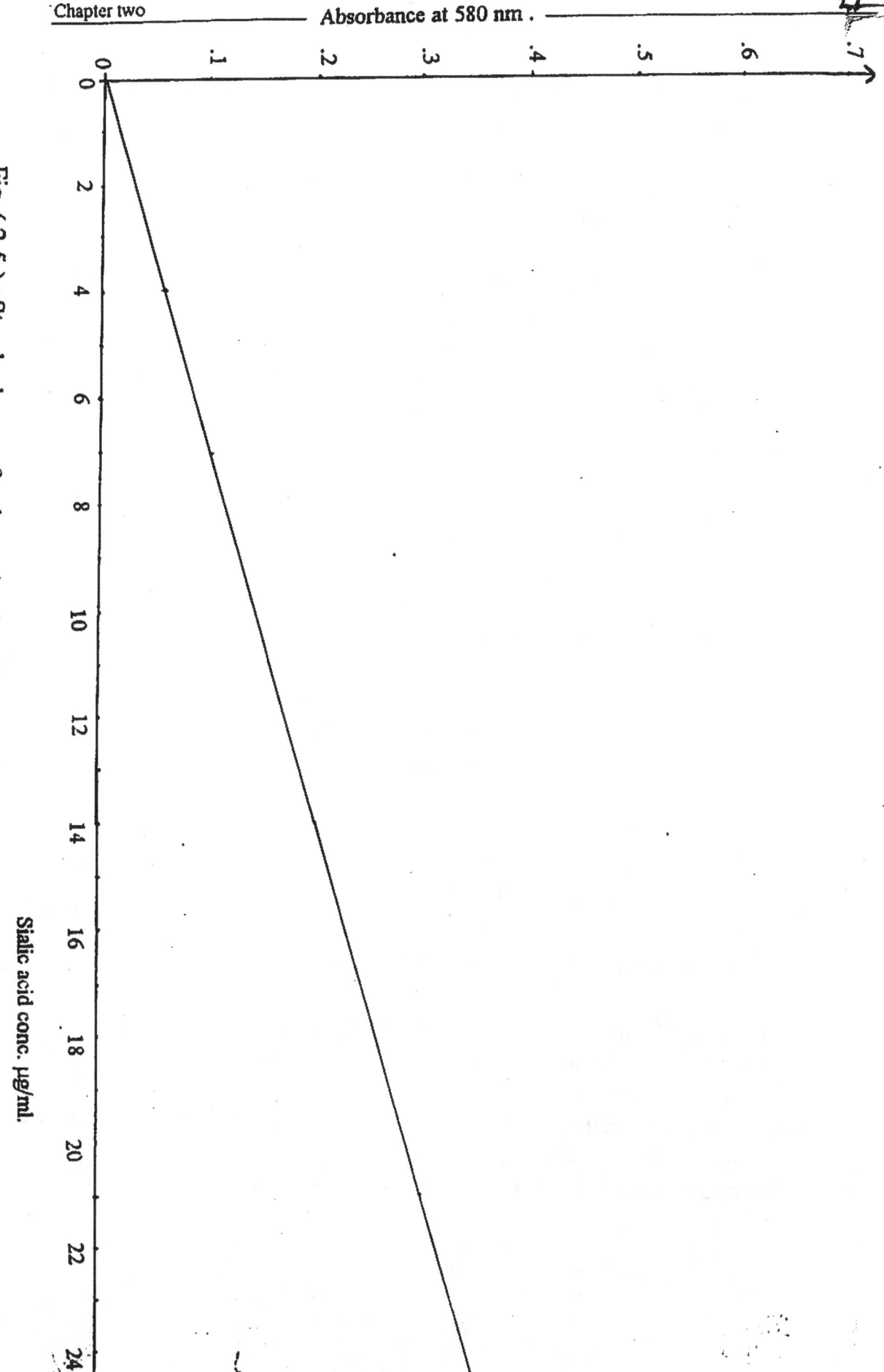

Fig. (2-5) : Standard curve for determination of sialic acid concentration.

2-8   Determination of seromucoid :

Seromucoid was determined according to Weimer and Mashin [21]. The method was carried out according to the following steps :-

1- 0.5 ml of serum was added to 4.5 ml of 0.85% NaCl with mixing , 2.5 ml of 1.8M $HClO_4$ was added and mixed by invertion , after 10 min it was centrifuged at 3000 r.p.m. to obtain supernatant .

2- To five ml of clear supernatant , 1 ml of phosphotungstic acid was added . Mixed and stand for 10 minutes , centrifuged , supernatant .

3- 0.5 ml of 0.1N NaOH was added to dissolve precipitate indicate as the unknown .

4- Adding 1.25 ml. orcinol reagent to unknown , blank and standard then 7.5 ml of 60% H_2SO_4 mix , place in 80±0.5℃ water bath for 20 minutes , cool , read aganist water and blank at 520nm .

Calculation :-

$$\text{Seromucoid (mg/100ml.)} = \frac{A_t - A_b}{A_s - A_b} \times 0.1 \times \frac{100}{0.333}$$

(expressed as galactose + mannose)

$$= \frac{A_t - A_b}{A_s - A_b} \times 30$$

While A_t = Unknown absorbance at 520 nm .

A_s = Standard absorbance at 520 nm .

A_b = Blank absorbance at 520 nm .

2-9   Determination of protein bound hexose :-

The method [153] should be followed according to the following steps :-

1- Adding 0.1 ml serum to 5 ml of 95% Ethanol . mix , centrifuged for 15 minutes , decant the upper layer .

2- Wash the precipitate with another 5 ml of 95% Ethanol . centrifuge and decant .

3- Follow the steps 3 - 4 in 2-8.

Chemical preparation were the same of seromucoid :-

Calculation :

$$\text{Protein bound hexose} = \frac{A_t - A_b}{A_s - A_b} \times 0.1 \times 100$$
(mg/100ml.)

$$= \frac{A_t - A_b}{A_s - A_b} \times 10$$

Where

A_t = Unknown absorbance at 520 nm .

A_s = Standard absorbance at 520 nm .

A_b = Blank absorbance at 520 nm .

Statistical analysis :-

Student's t-test was used for Statistical analysis of P values < 0.01 were considered significant.

2-10 Tissue collection of thyroid gland tumors :-

Surgery was carried out on patients (66,45,43 and 45 years of age) suffered from thyroid gland tumors (nodular goiter and multinodular goiter). in Medical city center and Al-Yarmok hospital. The upper scanning was done, and clinical diagnosis indicate enlargment of both lobes with multiple sizes. These tissues (non toxic nodular goiter and multinodular goiter) were immediately immersed in ice cold saline and then washed with phosphate - saline buffer pH 7.2 and kept at -20 °C until the homogenization process.

2-10-1 Preparation of human thyroid gland tumor homogenate :-

Three grams of thyroid tumor tissues were washed with 5 ml of 0.9% NaCl to remove surface mucus materials and contamination. then homogenized in 15 ml of 0.02M phosphate buffered saline 1:1 v/v (75mM KH_2PO_4 + 75 mM Na_2HPO_4) pH 7.2, containing 4 mM of 2-mercaptoethanol 2 mM EDTA and 0.075 M NaCl. using Tenbroeck ground - glass homogenizer to prpare the homogenate. The homogenate was centrifuged at 4000 r.p.m. about one hour. The supernatant was used as a source of crude lectin for binding (expressed as hemagglutination) studies . The supernatant was used through out study and stored at 20°C till using [154].

2-11 Determination of endogenous Ca^{++} in thyroid gland tumor homogenate :-

Endogenious Ca^{++} concentration of human thyroid gland tumor homogenate was determined by atomic absorption spectroscopy using $CaCl_2$ solution as standard figure (2-6).

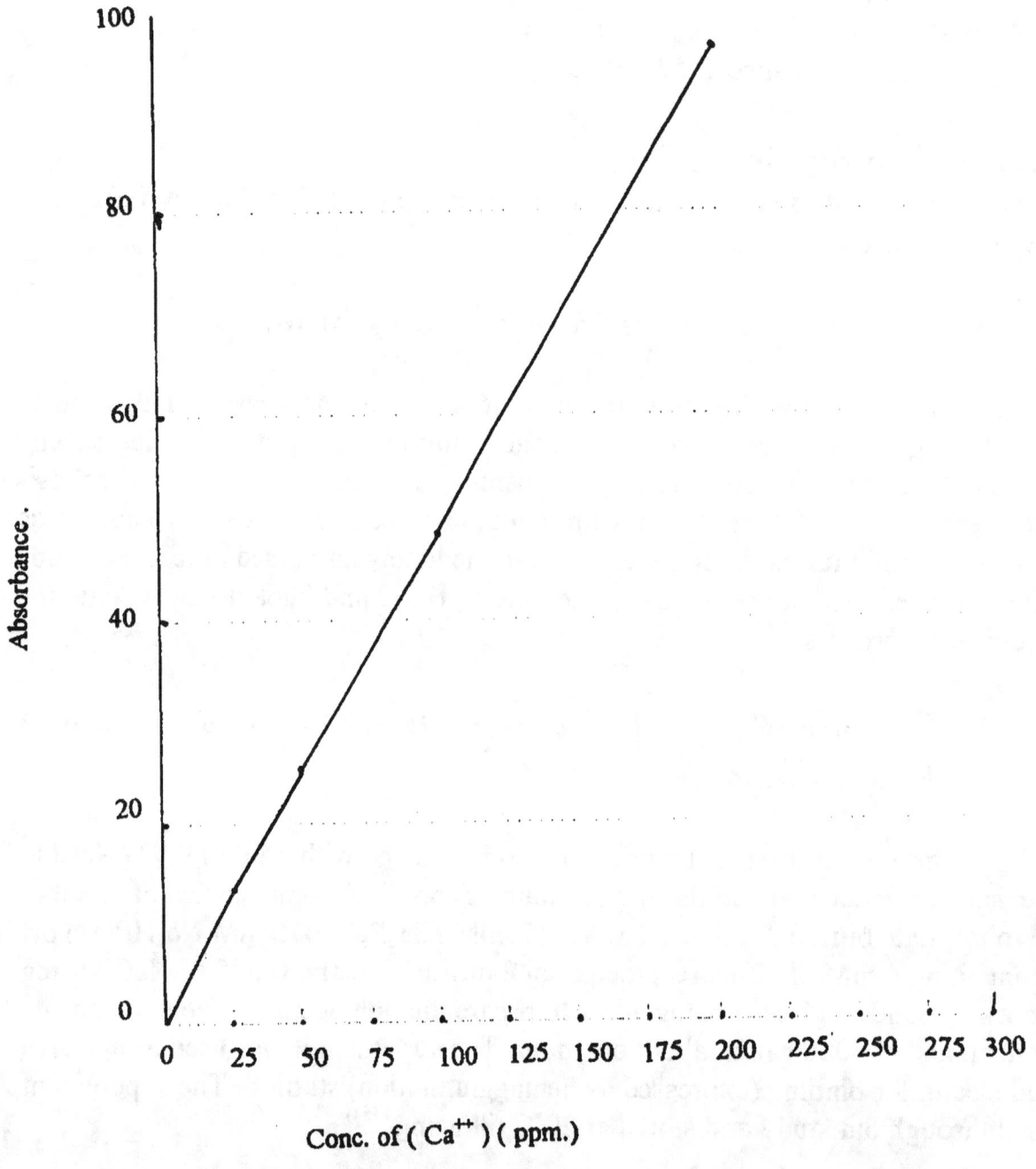

Fig. (2-6) : Standard curve for (Ca^{++}) determination in crude human thyroid gland lectin, using atomic absorption analysis .

2-12 Preliminary test for the binding of thyroid gland tumor lectin to erythrocyte suspension :-

2-12-1 The preparation of erythrocytes suspension for hemagglutination :-

Type A of human blood was obtained from blood bank . Taking 0.5 ml of whole blood then washed four times with 4 ml of 9% w/v NaCl and centrifuged for 3min. decantation of the upper layer was carried out. The suspension was diluted with 10 ml 0.9% w/v NaCl to give an absorbance of 2.000 at 620 nm.

2-12-2 The hemagglutinaation assay :-

The binding of thyroid tumor lectin to erythrocyte suspension was carried out by hemagglutination assay with modification of Liener method [155-156]. Half ml of diluted thyroid gland tumor homogenate was added to 2 ml of 0.02M Tris - Saline buffer pH = 9 then was incubated with 0.5 ml of erythrocyte suspension , at room temperature for 30 minutes . Then the cells was pelleted by centrifugation for 3min. at 3000 r.p.m. .The aggregated cell suspension solution were resuspend then allowed to stand for five minutes , the absorbance was determined at 620 nm of the upper half layer of the assay solution .

Calculation :-

Total binding expressed as % hemagglutinaation activity represents the amount of lectin which binds the erythrocytes and causing hemagglutination .

$$\% \text{ Hemagglutination activity} = \frac{A_{620} \text{ (red cells)} - A_{620} \text{ (red cells + Lectin)}}{A_{620} \text{ (red cells)}} \times 100$$

$$(\text{Total binding}) = \frac{A_{620} \text{ (Total red cells)} - A_{620} [\text{Total red cells -(complex)}]}{A_{620} \text{ (Total red cells)}} \times 100$$

2-12-3   Determination of non - specific binding of thyroid gland homogenate :-

The non - specific binding was determined by the same procedure except, the whole blood were washed four times with 0.9% w/v NaCl , four times with buffer (0.02M Tris - saline buffer), the suspension of red blood cells was prepared. Half ml. of this suspension placed in a test tube, incubated with various concentrations of glucouronic acid (0.5,1.0,1.5,2,2.5 and 3 mM), incubated for 30 min at 25°C. Then centrifuged at 3000 r.p.m. for 3 minutes , the aggregated cell suspension was mixed and allowed to stand for five minutes . The absorbance was determined at 620 nm of the upper half layer of the assay solution.

Calculation :-
Specific binding = Total binding - (non - specific binding)
   (S.B.)   (hemagglutination)
                      (activity)

$$\% \text{ S.B.} = \frac{\text{S.B.}}{\text{Total binding}} \times 100$$

2-12-4   Hemagglutination activity unit :

The hemagglutination activity unit is the level of test solution   (hence thyroid gland tumor homogenate concentration) which cause 50% of the standard cell suspension to sediment (0.5 - 2 hours) as determined by Lis and Sharon method[156] or by plotting the hemagglutination activity data vs. the lectin concentration .

2-12-5   Effect of pH on hemagglutination :-

Fifty µl of thyroid gland tumor homogenate was incubated with 2 ml of working buffer 0.02M Tris - saline buffer of various pH from (7 - 9) in presence of 0.001mM Ca^{++} for 30 minutes at room temperature, centrifuged at 3000 rpm for three minutes then resuspend and allowed to stand for five minutes, then determined the absorbance at 620 nm of the upper half layer of the assay solution .
Calculation :-
The same mathmatical equation was used in experiment 2-12-2 , to calculate the % of hemagglutination activity. The % of hemagglutination was plotted vs. their corresponding pH (7-9).

2-12-6   Effect of temperature on hemagglutin- ation activity :

Fifty μl of thyroid gland tumor homogenate was incubated with two ml. of working buffer 0.02M Tris - saline buffer pH 9, then half ml. of erythrocyte suspension was added for 30 minutes, at different temperature (6,15,20,25,30 &35°C). The final volume was 2.550 ml. then centrifuged for 3 minutes at 3000 r.p.m. the aggregated cell suspension was resuspend and allowed to stand for five minutes . The upper half layer was measured of the absorbance at 620 nm of the assay solution .

Calculation :-
The same mathematical equation mentioned in hemagglutination assay 2-12-2 was used to calculate the % of hemaggultion activity . The % of hemagglutination was plotted vs. their corresponding temperature .

2-12-7 The effect of incubation time on hemagglutination activity :-

Fifty μl of thyroid gland tumor homogenate was incubated with 2 ml of 0.02M Tris - saline buffer pH 9 and half ml. of erythrocyte suspension, at 25°C for varried incubation time (i.e. 0,15,30,45,60,75,90, 105,120,135and 150 min) then centrifuged for 3 minutes at 3000 r.p.m. the aggregated cell suspension was resuspend for five minutes, the absorbance was determined at 620 nm of the upper half layer of the assay solution.

Calculation :-
The same mathematical equation in experiment 2-12-2 was carried out to calculate the % of hemagglutination activity. The % of hemagglutination was plotted vs. their corresponding time.

2-12-8 Choice of the most appropriate amount of the human thyroid gland tumor homogenate:-

Half ml. of erythrocyte suspension was incbated with half mi. of various amounts of dissolved lectin homogenate in working buffer pH9 (78,156,234,312,390,468,546 and 624 μg). Two ml. of 0.02M Tris - saline buffer pH9 was added then in cubated for 30 min. at 25°C . Then cenrifugation for 3min. at 3000 rpm. Was carried out. The aggregated cell suspension was resuspend then allowed to stand for five minutes . The absorbance was determined at 620 nm of the upper half layer of the assay solution .

Calculation :-
The same mathematical equation in experiment 2-12-2 to was used calculate the % of hemagglutination activity (total binding) . The % of hemagglutination was plotted vs. their corresponding lectin amount .

2-12-9 Effect of exogenous Ca^{++} ion, on hemagglutination activity: -

Fifty μl of the human thyroid gland tumor homogenate was incubated with 0.5 ml of red cell suspension, 2 ml of 0.02M Tris - saline buffer pH = 9 containing various concentrations (2.5,5,10,15,20,25 and 30mM). The final reaction volume was 2.55 ml, centrifuged for 3 minutes at 3000 rpm . The aggregated cell suspension was resuspend and allowed to stand for five minutes , then the absorbance of the upper layer of the assay solution was read 620 nm of .

Calculation :-
The same mathematical equation in experiment 2-12-2 was used to calculate the % hemagglutination activity, then plotted vs. their corresponding Ca^{++} ion concentrations.

2-12-10 Effect of denaturating agents on the hemagglutination activity :-

Fifty μl of human thyroid gland tumor homogenate was incubated with two ml of working buffer 0.02M Tris - saline buffer pH 9 containing different concentrations of denaturing agents (3-6M Urea,0.5-4% polyethylene glycol),dissolved in Tris buffer the incubation time was one hour at 25°C, then 0.5 ml of red cell suspension was added, and the reaction was carried out as in hemagglutination assay section.

Calculation :-
Using the same mathematical formula in experiment 2-12-2 was used to calculate the % of hemagglutination activity .

2-12-11 Effect of ionic strength and different salts on hemagglutination activity :-

Effect of monovalent chloride salts on hemagglutination activity :-

Fifty μl of human thyroid gland homogenate in two ml. of 0.02M Tris - saline buffer pH 9, was incubated with 0.5 ml of erythrocyte suspension at 25°C for 16 min remembering that working buffer solution contains different concentrations of KCl and NaCl (0.005- 0.03M). The final volume was 2.550 ml, then the same steps was carried out as mentioned in 2-12-2.

Calculation :-
The same mathematical formula mentioned in experiment 2-12-2 was used to find out the % of hemagglutination activity . The % of hemagglutination was plotted vs. their corresponding monovalent salt concentration in the medium.

2-12-12 The effect of divalent chloride salts on hemagglutination activity :-

In this experiment twenty μl of human thyroid gland homogenate, diluted with 2.53 ml of 0.02M Tris - saline buffer pH 9, incubated with 0.5 ml erythrocyte suspension at 37°C, for 16 min. in the presence of 20mM $CaCl_2$ with various concentrations from 5×10^{-3} - 20×10^{-3} M of $MnCl_2$, $MgCl_2$ dissolved in the same buffer. The final volume 2.55 ml. and the reaction was carried out as in experiment 2-12-2.

Calculation :-
The same mathematical formula in the experiment 2-12-2 was used to find out the % of hemagglutination activity, the % of hemagglutination activity was plotted vs. the corresponding divalent salts concentrations.

2-12-13 Inhibition studies on binding of thyroid gland tumor homogenate and erythrocyte surface glycoprotein (hemagglutiation activity) :-

In this experiment different carbohydrates were (D-Fructose , D-galactose, D-mannose, D-xylose and sucrose) were used as inhibitors for human thyroid gland Lectin hemagglutination. The first step of this assay was carried out by high lectin concentration which give more than 90% hemagglutination .

The inhibition experiment have been done with D-glucouronic acid to find out the specific binding . The second step was carried out as in section 2-12-2. The percent of hemagglutination inhibition represents the difference between the % of hemagglutination activity, with lectin alone and that obtained with lectin plus inhibitor . Hence the assay have been carried out by taking 100 μl of thyroid gland tumor homogenate, diluted with 1.95 ml of 0.02M Tris - saline buffer pH 9 which contains the desired concentration of sugar (30mM) then incubated with 0.5 ml erythrocyte suspension at 25°C for 16 min . The reaction completed as in section 2-12-2. The % of Inhibition of hemagglutination was calculated according to mathematical formula section 2-12-2, then plotted vs. corresponding sugar concentration.

2-12-14 Effect of δ-globulin on hemagglutination activity:-

Various concentrations of δ- globulin by dissolving the appropriate amounts in the same working buffer was used in the assay , then the reaction carried out as in section 2-12-2 . The percent of inhibition of hemagglutination was calculated and plotted vs. their corresponding concentrations of δ- globulin.

2-13 Purification and identification of human thyroid tumor Lectin using formalinized erythrocytes and gel filtration:

2-13-1 Gel preparation and column packing :

The sephadx G-150 was allowed to swell in excess of 0.02M Tris buffer saline pH 7.2 containing 0.01M $CaCl_2$ and left to stand for over night at room temperature. Then the slurry was poured carefully into a vertical glass column down the wall using a glass - rod, the gel was settled, then the column (with the dimension of 1.5X75 cm) was equilibrated with 0.02M Tris - saline buffer pH 7.2 for 24 hours.

2-13-2 Void volume (V0) Determination :-

The volume of the gel was determined using blue dextran2000 at concentration 1mg/ml in 0.02M Tris - saline buffer pH 7.2 then after the addition of one ml of blue dextran solution to the column surface carefully, the elution carried out using the same buffer at flow rate of 8 ml/hour, fractions of three ml. were then collected and their absorbance was measured at 600 nm to determine the void volume (V_0) .

2-13-3 Column – Calibration

The column was calibrated by Kits purchased from Pharmacia Fine Chemicals that contain three standard proteins . Standard protein solution were prepared according to manufactures instructions, then applied through separate runs of each standard. The first run include albumin, the second run contain ovalbumin and the last run consist of ribonucleas A. Elution was carried out by 0.02M Tris - saline buffer with flow rate of 8 ml/hour . The collected fractions were absorbed at 280 nm to determine the elution volume (Ve) of each standard protein .

Calculation :-

1- The K_{av} values of the proteins eluted were calculated by the following equation :

$$K_{av} = \frac{V_e - V_0}{V_t - V_0}$$

while

V_0 = Void volume

Ve = Elution volume of each protein

Vt = total gel - bed volume (determined from the following equation :-

$Vt = (\frac{column\ diameter}{2})^2 \times \frac{22}{7} \times column\ length$

2- Calibration curve was plotted .

$[(- \log K_{av})^{1/2}]$vs. stocks radius (Rs) of the proteins [163]

2-13-4   Steps of purification :

1- Preparation of formalinized erythrocytes as affinity matrix: -

The essential methods as described by Csizman 1960 [157] and Nowak and Borondes 1975[158] was used for preparation of formalinized erythrocytes. These human erythrocytes (type A) were supplied by blood bank, washed four times in twenty volume of 75mM NaCl, 75mM Na_2HPO_4 KH_2PO_4 (phosphate - saline buffer pH 7.2 per packed cell volume by centrifugation at 300 r.p.m. for ten minutes. Resuspention of twenty five ml of the packed red cells in to 200 ml of phosphate - saline buffer pH 7.2 and placed in a 500 ml conical flask. Fifty ml. of commercial formalin (40% formaldehyde) was introduced and the mixture was incubated at 37°C for 20 hours. The cells then washed five times in five volumes phosphate - saline buffer pH 7.2 per packed cell volume and kept at 4°C.

The stored formalinized cells prepared were used as an affinity reagent by washing the cells six times in ten volumes of 50 mM Tris - HCl, 100 mM NaCl (Tris-saline buffer) pH 7.2. Ten ml of thyroid gland tumor homogenate (21.923mg/ml protein) was added to thirty ml of cell suspension (10% suspension in buffer) for three hours at room temperature . then washed three times by twenty volumes of Tris - saline buffer pH 7.2 . The elution of the adsorbed lectin was accompanied by incubation the cell with fifty ml of 0.15M D-glucouronic acid in Tris - saline buffer which has been brought to be 7.2 with NaOH before the addition to the cells and placed in refrigerator for overnight .The elution mixture was centrifuged for ten minutes at 3000 r.p.m. and the resultant supernatant was referred as fraction no. 2 , whereas the thyroid gland homogenate being fraction no. 1 .

2- Gel Filtration:

The column had been equilibrated with Tris saline buffer pH 7.2, the sample was transferred at the top surface of the sephadex G 150 column (1.5 cm X 75 cm), then eluted with this buffer at on elution rate 8 ml/hour , the fraction of three ml. volume were collected then identified by the assay method as well as the absorbance at 280 nm and protein determination were carried out . The elution volume (Ve) of thyroid gland tumor lectin in each fraction was determined by the following formula:-

Ve = Fraction volume (3 ml) X Fraction number containing the highest level of thyroid gland tumor lectin .

3- The assay method:-

In order to identify the fractions containing lectin , the hemagglutination (binding) was determined as follows :-
Taking half ml of each fraction isolated by gel filtration then was incubated with 0.5 ml of red cells suspension in a final volume 2.55 ml. at 25°C for 16 minutes, noticing that parallel experiments was performed to determine the amount of non - specific binding for each fraction .
Calculation:
Using the following mathematical formula to calculate the % of hemagglutination activity of purified thyroid gland tumor lectin :-

$$\% \ Hemagglutination = \frac{A_{620} (red\ cells) - A_{620} ((red\ cells) - purified\ lectin)}{A_{620} (red\ cells)} \times 100$$

2-13-5 Determination of purified fold

The purification fold for the lectin was determined using the following formula:

$$Purification\ fold\ of\ lectin = \frac{Specific\ activity\ of\ pure\ lectin}{Specific\ activity\ of\ the\ homogenate\ lectin}$$

2-13-6 Determination of % Recovery of lectin :

The % of recovery was determined as follows:

$$\% \text{ Recovery} = \frac{\text{Total unit of pure lectin form}}{\text{Total units of homogenate}} \times 100$$

2-13-7 Analysis of the purified fraction by acrylamide gel – electrophoresis (PAGE) :-

It was performed by gel electrophoresis in the presence of sodium dodecyl sulphate (SDS) according to the method of Laemmli [159], using 7.5% Acrylamide separating gels . To determine the molecular weight of thyroid gland tumor lectin. Known molecular weights of standard proteins were used. Phosphorylase b (94000), BSA (67000), ova Albumin (43000), Carbonic anhydrase (30000), Trypsin (20000) and α- Lactoalbumin 14000 .

Procedure :

1- Polyacrylamide gel (7.5% concentration) was prepared according to the application note 306 issued by LKB company.

2- Standard proteins solutions:
Pharmacia electrophoresis calibration kit for the determination of native molecular weight of proteins by polyacrylamide - gel electrophoresis was used[160]. The content of each vial was redissolved in 0.3 ml of sample buffer (0.0625M Tris - HC l pH 6.8 containing 2% SDS , 10% glycerol , 5% 2 - mercapto ethanol and 0.001% bromophenol blue as the dye) . These standard proteins was used phosphorylase b 94000 , Albumin 67000 , ova albumin 43000, Carbonic anhydrase 30000 , Trypsin inhibitor 20000 and α- Lactoalbumn 14000.

3- Sample preparation :
Purified lectin was concentrated by dialysis against sucrose , then diluted with sample buffer to protein concentration range (0.2 - 2) mg/ml. The other steps which include fixation , staining . destaining and preserving were carried out as explained by S. A. , North M and Colson [15] 1987 and application note 306 of LKB company (for voltage and current used for separation note 306 [160]) .

Calculation:-

1- The relative mobility Rm of each protein was measured as follows :

$$Rm = \frac{\text{distance of protein migration}}{\text{length after dying}} \times \frac{\text{length before fixation}}{\text{distance of dye migration}}$$

2- Log Mol. Wt. of the marker standard proteins was plotted against relative mobilites (Rm) , typically gives a straight line Figure (3-18) from which the purified lectin molecular weight was estimated .

2-13-8 Determination of molecular weight of lectin using Kav. values :-

The K_{av} value for the lectin was determined using the following formula:

$$K_{av} = \frac{Ve - V_0}{VT - V_0}$$

K_{av} value of thyroid gland tumor lectin was applied to the calibration curve vs. standard molecular weights to determine its molecular weight .

2-13-9 Determination of lectin stock radius :-

The value $(- Log\ Kav)^{1/2}$ of standard proteins were determined and applied to stocks radius (Rs) in the curve figure (3-19) to determine the lectin stock radius.

2-14 Kinetic studies of human thyroid gland lectin :-

2-14-1 The time – course of binding of thyroid gland tumor lectin to its glycoprotein present on red blood cell surface:-

1- At zero time 50 µl of thyroid tumor lectin were incubated with 0.5 ml erythrocyte suspension. The final volume was made up to 2.550 ml by adding the assay buffer (0.02M Tris - buffer pH 9). Then incubated at 25°C for several time intervals (2,4,8,12,16,18 minutes) respectively.

2- After each time interval the assay tube was taken to determine the percent of hemagglutination .

3- Parallel experiments were performed to determine specific binding (as expressed in hemagglutination assay).

Calculation :-

   The value of lectin bound specifically was calculated according to the following formula :-

The value of specifically bound lectin

$$\text{The value of specifically bound lectin} = \frac{A_{620}\ \text{red cells} - A_{620}\ (\text{red cells} + \text{lectin})}{A_{620}\ \text{red cells}} \times \text{Lectin conc. (M)}$$

   The value of specific binding lectin were plotted vs. the incubation time .

2-14-2 Scatchard Analysis :

   Incubate 0.5 ml. erythrocyte suspension with increasing concentration amounts of thyroid gland lectin .

   The final volume was made up to 2.550 ml by adding 0.02M Tris - saline buffer pH 9 containing $CaCl_2$, the experiment was carried out in triplicate , then incubate the tubes at 25°C for 16 minutes .

Calculations: -

   Let B is the bound lectin with red blood cell surface (complex). and F is the free form of lectin (unreacted lectin). The value of bound lectin concentration in molar was calculated using the following formula:-

$$B = \frac{A_{620} \text{ of red cell} - A_{620} \text{ (red cell + lectin)}}{A_{620} \text{ red cell}} \times \text{ lectin conc. (M)}$$

The B value was plotted vs. lectin concentration (M) used and the plot of B/F value vs. the B specific gives linear relationship. The value of affinity constant of the binding at each temperature can be calculated from the slop of the straight line while the value of the total binding site concentration or number (B max) was calculated from the relation

$$\frac{B}{F} = \frac{1}{K_d} (B_{max} - B_{specific})$$

$$B = - K_d \frac{[B]}{[F]} + B \text{ max}$$

So the value -Kd is equal to the slope of the curve and B max value was obtained from the intercept on ordinate.

2-14-3 Determinatin of Hill coefficient (n) of lectin to glycoprotein.

1- The informational data was obtained from section (2-14-2).
2- The Hill - coefficient were obtained using the following equation, known as Hill - equation.

$$\text{Log} [B \text{ spec.} / B \text{ max} - B \text{ spec}] = n \log L - \text{Log } K_d$$

where L = Free lectin in the incubation medium.
3- Then plot B spec. / B max - B spec. vs. Log L. concentration. The slop of the straight gives the Hill coefficient (n) value.

2-15 The thermodynamic studies :

1- Fifty μl. thyroid gland tumor was incubated with 0.5 ml of erythrocyte suspension. The total volume was 2.550 ml. by adding Tris - saline buffer pH 9. The incubation time was 16 min. at different temperatures 6,15,20 and 25°C. respectively. The absorbance was readed at 620 nm. and the B specific was calculated by the following equation :-

$$B \text{ specific} = \frac{A_{620} \text{ (red cells)} - A_{620} \text{ (red cells + lectin)}}{A_{620} \text{ (red cells)}} \times \text{Lectin conc.(M)}$$

2- The thermodynamic parameters of standard state were obtained from Van't Hof plot, the value of the Log Keq at different temperatures were plotted vs. the reciprocal values of absolute temperature in Kelvin ($\underline{1}$), according to the following equation :   T

$$\ln Keq = \Delta S° / R - \Delta H° / RT$$
Where :
    $\Delta H°$ = the enthalpy change of the standard state .
    $\Delta S°$ = the entropy change of the standard state .
      R = gas constant = $8.341 \ J \ K^{-1}$
The change in Gibbs free energy of the standard state ($\Delta G°$) was obtained from the following equation :-

$$\Delta G° = - RT \ln Ka$$
While the standard state entropy change was obtained as follows :-

$$\Delta S° = [\Delta H° - \Delta G°] / T .$$
3- The thermodynamic parameters of transition state were obtained from Arrhenius plot of $\ln k_{+1}$ values vs. $\underline{1}$ gives a linear relationship obtained from the following equation :   T

$$\ln k_{+1} = \ln A - Eq / RT$$
While A = Arrhenius constant or frequency factor .
The apparent energy value of activation Eq of the binding reaction was obtained from the slop the straight line. The enthalpy of transition state ΔH^* can be calculated from the following equation:-

$$\Delta H^* = Ea - RT$$
Transition state free energy change is found by using this equation :-

$$\Delta G^* = - RT \ln K_{+1} + RT \ln \frac{KT}{h}$$

where k = $1.3806 X \ 10^{-23} \ JK^-$ Boltzman's constant
     h = $6.626 X 10^{-34}$ JS Planck's constant
The change in entropy of the transition state ΔS^* was found from the following :-

$$\Delta S^* = [\Delta H^* - \Delta G^*] / T .$$

Results And Discussion

3-1 Estimation of T3, T4 and TSH in sera of thyroid gland tumor diseases and thyroid scanning :-

Thyroid scanning:-

The concentrating property of thyroid was used to estimate thyroid function and to obtain functional anatomical images of the gland. For these studies tracer amounts of Technetium - 99m administered to the patient. Gamma rays emitted by Technetium - 99m concentrated in the thyroid were detected by specially designed and counting devices, then transformed into thyroid technetium - 99m uptake and images. Figure 3.1 (a - g), which show the response of different thyroid glands during scanning. These thyroid glands were estimated functionally through anatomical images of these glands. This technique is successfully used to diagnose the diseases chosen in this work such as thyroid carcinoma, multinodular goiter, nodular goiter, diffused goiter and normal thyroid gland. Laboratory findings for Technetium - 99m isotope localization may provide evidence for areas of altered activity as in active (tumor) and hyperactive in thyroid gland tissues. nodular lesion, multinodular lesion, Figure 3.1: (c - g).

The differentiated carcinoma may be correlated with the size determination carried out by scanning and thyroid lesion or thyroid enlargement by the scintillation camera dynamic scan .

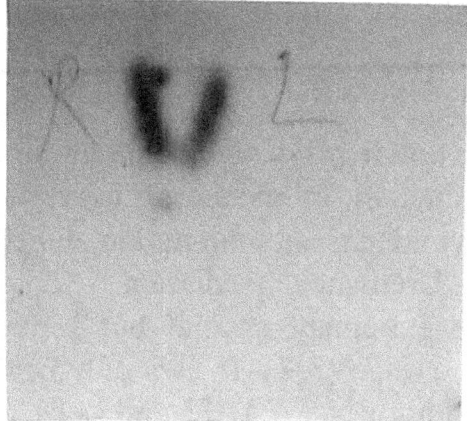

a- Normal human thyroid gland .

b- Thyroid carcinoma .

c- Thyroid carcinoma .

d- Multi nodular goiter .

Fig. 3.1 (a-g) : The Scanning of thyroid gland . All details are explained in text .

3-1-2   T3 , T4 and TSH levels :-

Patients with thyroid cancer usually do not displays significant abnormalities of thyroid function, so thyroid hormone production by metastasis is very rare [161] due to the environmental factors causing thyroid carcinoma, iodine deficiency, chemical carcinogen: (2-acytyl amino fluorene, Urethane), Irradiation [162-163]. Serum levels for L-T3 ($p<0.005$),L-T4 ($p<0.1$) are extremely normal but serum TSH ($p<0.0005$) was elevated as in table (3.1) and Fig. (3-2) In sera of hyperthyroid decreased levels of TSH and elevation of thyroid hormone (L-T3 and L-T4) were obtained These levels of L-T3 and L-T4 were significant ($p<0.0005$),while TSH level was of ($p<0.05$).

Serum TSH level are raised in cases of hypothyroidism. The diagnosis of hypothyroidism by the finding of a low T4 value was confirmed by a raised TSH level and thyroid scanning. A clinical state of this disease develops whenever insufficient amounts of thyroid hormone available to the tissue. The advanced case appear with high TSH ($p<0.05$) and low T4 ($p<0.1$) and T3 of ($p<0.1$) levels are shown also in table (3.1) and Fig(3-2).

Other samples chosen from patients after many years of thyroidectomy characterized by elevation of L-T3 and L-T4 , suppression of TSH were also assessed. These results attributed to differentiated carcinoma recurrence .

Table (3-1) : Laboratory findings for thyroid function test in various clinical conditions of thyroid gland patients :-

Pathological conditions	T3	T4	TSH	Thyroid scan
Normal	N	N	N	Absorbance of Tc $_{.99m}$ by normal thyroid gland gives normal shape of thyroid configuration
Hypothyroidism	↓	↓	↑,N	
Hyperthyroidism	↑	↑	↓,U	This type of tumor dependent on Tc $_{.99m}$ up take to obtain nodular and multinodular hyperthy roidism .
Nodular goiter	↑	↑	↓,U	
multinodular goiter	↑	↑	↓,U	
T$_3$ Toxic goiter	↑	↑	↓	
T$_4$ Toxic goiter	N	↑	↓	
Thyroid carcinoma	N	N	↑	The thyroid gland appear to have small location of Tc $_{.99m}$ uptake but the rest of the gland with out Tc $_{.99m}$ uptake .

While :-

N = Normal,   U = undetectable,   ↑ = increased,   ↓ = decreased.

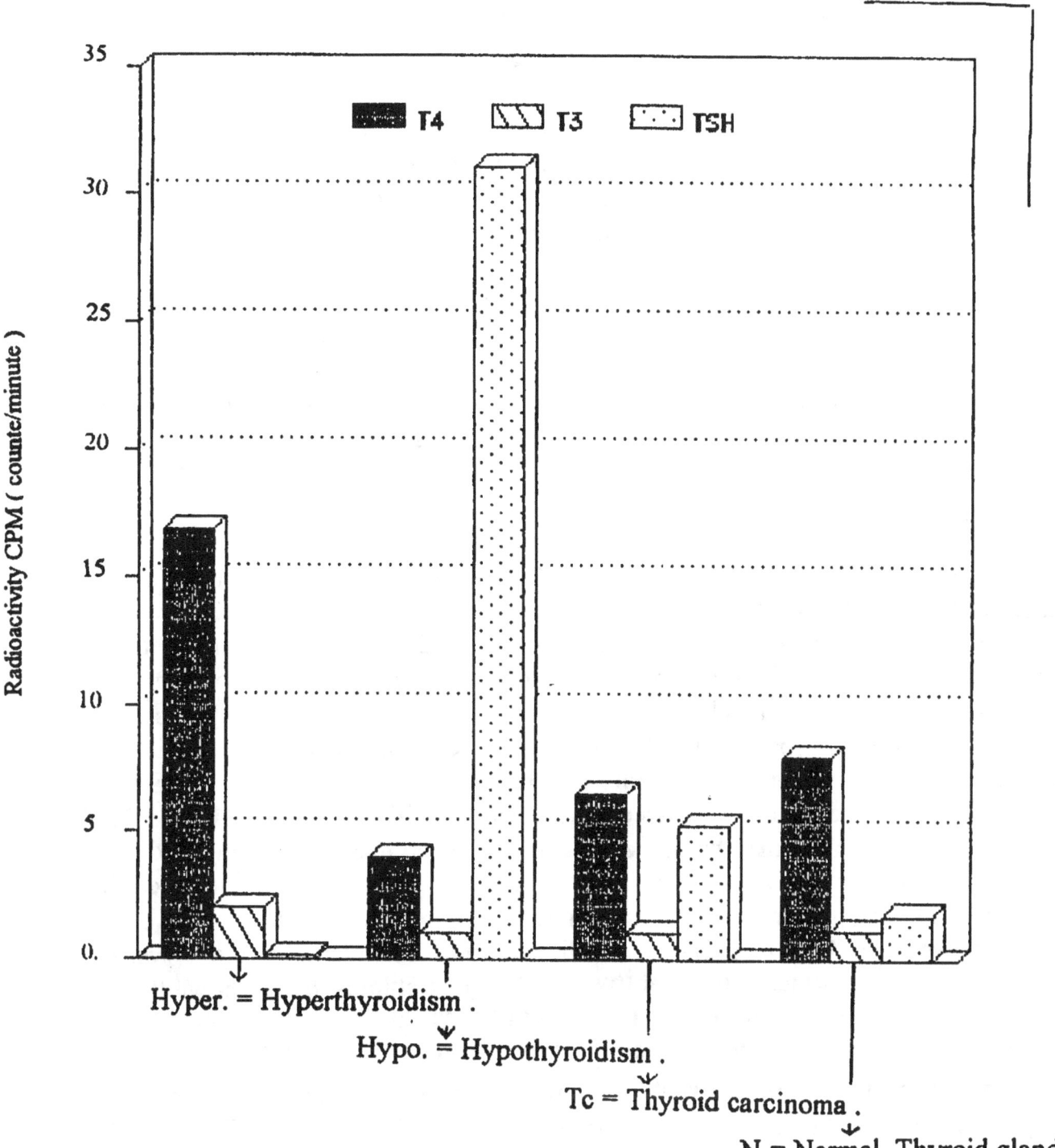

Fig. (3-2) : Determination of T_4 , T_3 & TSH in various thyroid diseases .

3-2 Determination of total sialic acid and lipid associated sialic acid in sera of thyroid gland tumor diseases .

Table (3-2) and figures (3-3) and (3-4) represents the comparison of the mean of TSA , LASA and TP values from sera of different cases of hyperthyroidism such as nodular goiter, multinodular goiter, goiter and toxic goiter, hypothyroidism, thyroidectomy, thyroid carcinoma , pathological control and normal healthy control . The TSA levels from different types of thyroid patients were significantly increased (p<0.001), when compared to normal healthy controls. Similar results were obtained with LASA level in both thyroid diseases and normal healthy controls . LASA level of the thyroid carcinoma and breast carcinoma , were also studied . The thyroid carcinoma patients show high level of LASA level compared to healthy individuals . Comparison of serum LASA level of all pathological cases were also studied . The thyroid carcinoma patients was also compared to NG , MNG , diffused goiter and other thyroid diseases .

The Tables (3-3 : A-D) support these finding statistically total sialic acid level was increased in sera of nodular , multinodular goiter , toxic goiter , diffused goiter , thyroid carcinoma , hypothyroidism and breast carcinoma , while the level in sera collected from the patient underwent thyroidectmoy directly remains unaffected . On the contrary ; the level in the sera collected after years from thyroidectomy was increased significantly. Total sialic acid level to total protein (TSA/TP) showed a significant increases (p<0.0005) for all type of diseases used Table (3.2) when compared to normal healthy controls. There is significant increase in LASA , TP . TSA/TP in most diseases compared to normal healthy controls table (3.2) . Similar results was obtained by Mrochek et al [164] when measuring sialic acid level in women with breast cancer. Increasing level of sialic acid in progressive cancer but a decrease in the levels in patients responding to therapy was obtained [165] . The mean of TSA was significantly higher in malignant melanoma and cardio [167]vasicular mortality [166] [132], benign tumors of thyroid [168], other studies have shown that TSA and LASA were elevated in cancer diseases and hence the TSA was considered as a tumor marker for monitoring cancer, compared with normal values [169-170].

The results of this work showed that LASA appeared to be good indicator for thyroid carcinoma , than TSA , due to the increased level in all thyroid tumors [171-172]
.

These results are more appropriate and specific for diagnosis of thyroid tumor in human sera, or it could be used as a tumor marker in these cases.

The level of TSA in NG and MNG was elevated less than that of thyroid carcinoma. Although thyroidectomy sera have shown an elevation directly but a increase in the level was clear after several years after thyroidectomy. The results then ascertained that sialic acid has an important role in tumor cells.

Specificity and sensitivity tests of TSA and LASA are mentioned in tables (3-4, 3-5) using 65 mg/100 ml. and 17 mg/100ml. for TSA , LASA as the upper limit of normal receptively. Specificity was calculated and found to be 95-100% for the normal range of TSA, so that 5% of normal test are falsely positive. Sensitivity and specificity of the LASA test was calculated using 17 mg/100 ml as the upper limit of normal. Sensitivity in cancer varied from 42 - 80% but the specificity measurement show 100% of the cases tested were falsely positive. LASA sensitivity test shows 20-100% for thyroid tumors and 40% for normal controls. Hence the sensitivity was greatly increased and specificity was increased also.

The determination of TSA and LASA was used to describe the changes that occur on the surface of different thyroid tumor cells and the ratio TSA /TP reflect that, total sialic acid may be more appropriate than LASA for assessment of thyroid tumors. On the contrary, the direct thyroidectomy sera showing normal values of TSA and LASA while 9 years after thyroidectomy have given elevated means of these parameters.

In spite of these findings correlation was calculated as in table (3-4), to establish the relationship between both (T3, T4 and TSH level) with total sialic acid level .

The correlation factor found to be positive for L-T4 levels . TSA levels in hyperthyrodism , and positive correlation for TSH levels with TSA levels in both hypothyroidism and thyroid carcinoma was obtained tables (3-6) and (3-7) .

Table(3-2): Comparison of total sialic acid (TSA) , lipid associated sialic acid (LASA) and total protein (TP) values for patients with hyperthyroidism. hypothyroidism thyroid carcinoma and normal group (mean±S.D.) .

Group	No. of cases	TP g/dl	TSA mg/dl	LASA mg/dl	TSA/TP mg/g
· Normal control	20	6.7±0.3	55.5±9.425	11.2±3.3	8.3±0.85
- Hyperthyroidism:- nodular goiter.	15	7.203±.0.318	85.7±7.17	14.8±2.56	11.95±0.566
multinodular goiter.	14	7.972±0.747	82.36±6.35	17.61±2.113	11.02±0.73
Toxic goiter	25	8.54±0.456	121.04±8.8	25.73±3.96	16.34±0.79
Diffused goiter	12	7.743±0.769	101.5±5.5	18.78±2.73	13.07±0.943
·· Hypothyroidism	7	7.735±0.598	134.6±8.63	19.39±2.46	17.4±0.952
··Thyroid carcinoma	12	7.633±0.374	128.75±8.9	27.25±6.196	17.64±0.89
·· Years after thyroidectomy	5	5.9176±0.62	118.8±7.3	25.33±3.32	18.715±0.534
·· Directly after thyroidectomy	5	7.1±0.35	50.6±9.75	12.96±2.25	13.07±0.62
· Other pathological conditions (Breast cancer)	10	7.65±0.569	127.5±10.88	26.5±4.74	16.8±0.866

Table (3-3,A) : Statistical evaluation of total serum protein levels g/dl ±SD , in sera of thyroid gland tumor diseases :-

Group	No. of cases	Std. Deviation	Variant (Std.)^2	Average.	Std. Error	Critical value of t-test	Significance value (P value).
Hyperthyroidism Nodular goiter	15	0.318	0.101	7.203	0.082	1.58	0.05
Multi nodular goiter	14	0.747	0.558	7.972	0.1996	1.703	0.05
Toxic goiter	25	0.456	0.208	8.54	0.0912	4.035	0.0005
Goiter	12	0.769	0.591	7.743	0.222	1.356	0.1
Hypothyroidism	7	0.598	0.358	7.735	0.226	1.7308	0.05
Thyroid carcinoma	12	0.374	0.1398	7.633	0.108	2.495	0.01
Years after thyroidectomy	5	0.35	0.1225	5.9176	0.1265	2.235	0.05
Directly after thyroidectomy	5	0.83	0.689	7.1	0.371	0.4815	0.1
Pathological controls (Breast cancer)	10	0.569	0.324	7.65	0.1799	1.6696	0.05
Normalvalues	20	0.3	0.09	6.7	0.067		

Table (3-3 , B) : Statistical evaluation of total sialic acid levels mg/100 ml. ± SD in sera of thyroid gland tumor diseases :-

Group	No. of cases	Std. Deviation	Variant (Std.)^2	Average.	Std. Error	Critical value of t-test	Significance value (P value).
Hyperthyroidism: Nodular goiter	15	7.17	51.41	85.7	1.85	4.21	0.0005
Multinodular goiter	14	6.35	40.32	82.36	1.697	4.23	0.0005
Toxic goiter	25	8.8	77.44	121.04	1.76	7.45	0.0005
goiter	12	5.5	30.25	101.5	1.588	8.36	0.0005
Hypothyriodism	7	8.63	214.04	134.7	3.26	9.17	0.0005
Thyroid carcinoma	12	8.9	79.21	128.75	2.57	8.23	0.0005
Years after thyroidectomy	5	7.3	53.29	118.8	3.27	8.67	0.0005
Directly after thyroidectomy	5	9.75	95.06	50.6	4.36	0.503	0.1
Pathological controls (Breast cancer)	10	10.88	118.37	123.5	3.44	6.62	0.005
Normal values	20	9.425	88.83	55.5	2.107		

Table (3-3 . C) : Statistical evaluation of total sialic acid / total protein ratio mg/g ±SD in sera of thyroid gland tumor diseases :-

Group	No. of cases	Std. Deviation	Variant (Std.)²	Average.	Std. Error	Critical value of t-test	Significance value (P value).
Hyperthyroidism: Nodular goiter	15	0.566	0.32	11.95	0.146	6.448	0.0005
Multinodular goiter	14	0.73	0.533	11.02	0.195	3.726	0.005
Toxic goiter	25	0.79	0.624	16.35	0.185	10.19	0.0005
Goiter	12	0.943	0.89	13.07	0.272	5.058	0.0005
Hypothyroidism	7	0.592	0.35	17.4	0.224	15.35	0.0005
Thyroid carcinoma	12	0.89	0.79	17.64	0.257	10.49	0.0005
Years after thyroidectomy	5	0.534	0.285	18.715	0.243	19.51	0.0005
Directly after thyroidectomy	5	0.62	0.384	13.07	0.277	7.694	0.005
Pathological controls (Breast cancer)	10	0.866	0.75	16.8	0.739	9.815	0.0005
Normal values	20	0.85	0.723	8.3	0.190		

Table (3-3. D): Statistical evaluation of lipid associated sialic acid level mg/dl ±SD in sera of thyroid gland tumor diseases :-

Group	No. of cases	Std. Deviation	Variant (Std.)²	Average.	Std. Error	Critical value of t-test	Significance value (P value).
Hyperthyroidis Nodular goiter	15	2.56	6.554	14.8	0.661	1.406	0.1
Multinodular goiter	14	2.113	4.465	17.61	0.565	3.034	0.005
Toxic goiter	25	3.96	15.68	25.73	0.792	3.669	0.0005
Goiter	12	2.73	7.45	18.78	0.788	2.78	0.01
Hypothyroidism	7	2.46	6.052	19.39	0.93	3.331	0.01
Thyroid carcinoma	12	6.196	38.39	27.25	1.789	2.59	0.01
Years after thyroidectomy	5	3.32	11.02	25.33	1.485	4.256	0.005
Directly after thyroidectomy	5	2.25	5.063	12.96	2.264	0.782	0.1
Pathological controls (Breast cancer)	10	4.74	22.47	26.5	1.499	3.228	0.01
Normal values	20	3.3	10.89	11.2	0.738		

Table(3-4): Specificity of TSA and LASA tests . Calculations were carried out by taking the number of cases which have 65 mg/ml or below and divided by total number of cases .

TSA test Specificity	No. of cases	TSA≤65 mg/dl (normal)	specificity true negative	LASA≤17 mg/dl (normal)	specificity true negative
Hyperthyroidism: Nodular goiter	15	1	6.66	12	80
Multinodular goiter	14	2	14.28	6	42.86
Toxic goiter	20	0	0	0	0
goiter	12	0	0	7	58.33
Hypothyroidism	7	0	0	0	0
Thyroid carcinoma	12	0	0	0	0
Directly after thyroidecomy	5	1	20	5	100
Years after thyroidectomy	5	0	0	0	0
Pathological control (Breast cancer)	10	0	0	0	0
Normal	20	19	95	20	100

Table(3-5): Sensitivity of TSA and LASA tests. Calculations were carried out by taking the number of cases which have ≥ 65 mg/dl values and divided by total number of cases .

TSA test sensitivity	No. of cases	TSA≥65 mg/dl TSA elevated	Sensitivity negative	LASA≥17 mg/dl LASA elevated true	Sensitivity negative
Hyperthyroidism: Nodular goiter	15	14	93.33	3	20
Multinodular goiter	14	12	85.71	8	57.14
Toxic goiter	20	20	100	20	100
Goiter	12	12	100	5	41.66
Hypothyroidism	7	7	100	7	100
Thyroid carcinoma	12	12	100	12	100
Directly after thyroidectomy	5	0	0	0	0
Years after thyroidectomy	5	5	100	5	100
Pathological control (Breast cancer)	10	10	100	10	100
Normal	20	3	15	8	40

Table (3-6): Correlation factor between TSA and L-T3 , L-T4 and TSH in serum from thyroid tumor patients :

Diagnosis	Correlation (r) of TSA with :-		
	L-T4	L-T3	TSH
Hypothyroidism	- 0.0778	0.04	zero
Hyperthyroidism	0.06816	0.02	0.545
Thyroid carcinoma	$- 2.244 \times 10^{-4}$	zero	0.172

Table (3-7) : L-T$_3$, L-T$_4$,   TSH and TSA in sera of thyroid tumor patients .

Diagnosis	Number	L-T$_4$ μg/100 ml	L-T$_3$ ng/ml	L-TSH μIU/L	TSA mg/100ml
Normal	20	7.99±0.691	1.159±0.0132	1.796±0.137	55.5±9.425
Hyperthyroidism	20	16.893±2.89	2.097±0.105	0.194±0.08	121.04±8.8
Hypothyroidism	16	4.04±2.069	1.118±0.0558	31.07±24.774	134.6±4.624
Thyroid carcinoma	11	6.55±1.65	1.12±1.01	5.33±0.235	128.75±4.624

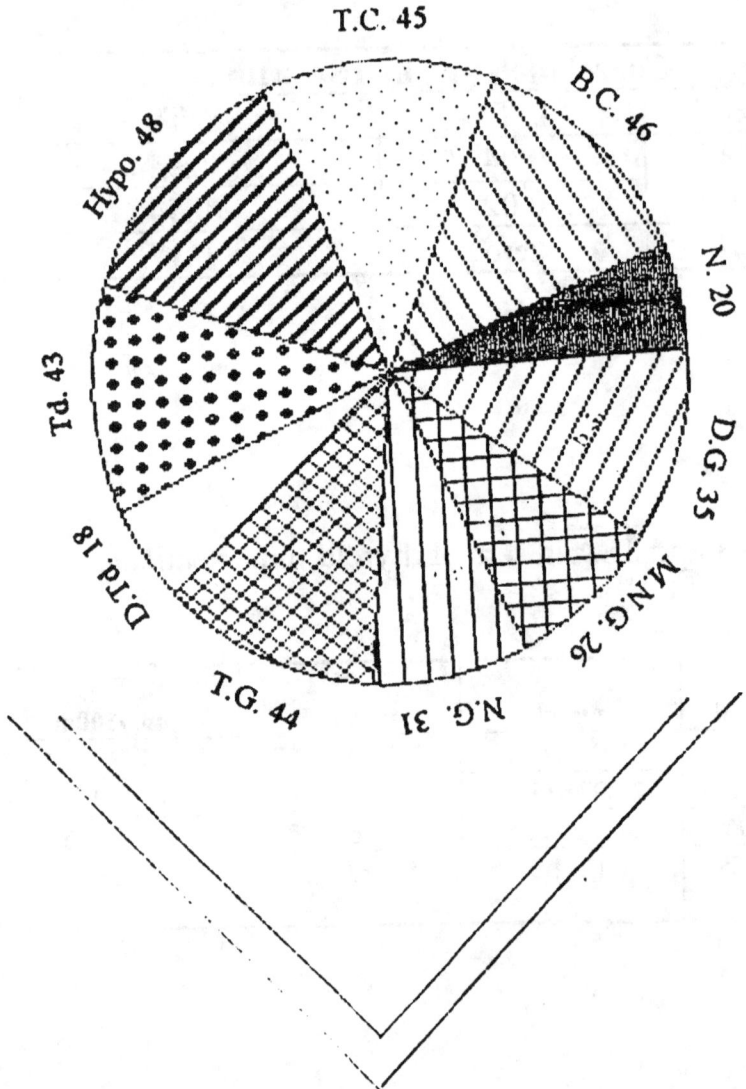

Fig. (3-3) : Distribution of TSA levels in sera of normal volunteers (N), pathological controls (BC.), thyroid carcinoma (TC.), hypothyroidism (Hypo.), years after thyroidectomy (TD), directly after thyroidactomy (D,TD) toxic goiter (TG), nodular goiter (NG), multinodular goiter (MNG), and goiter (D.G).

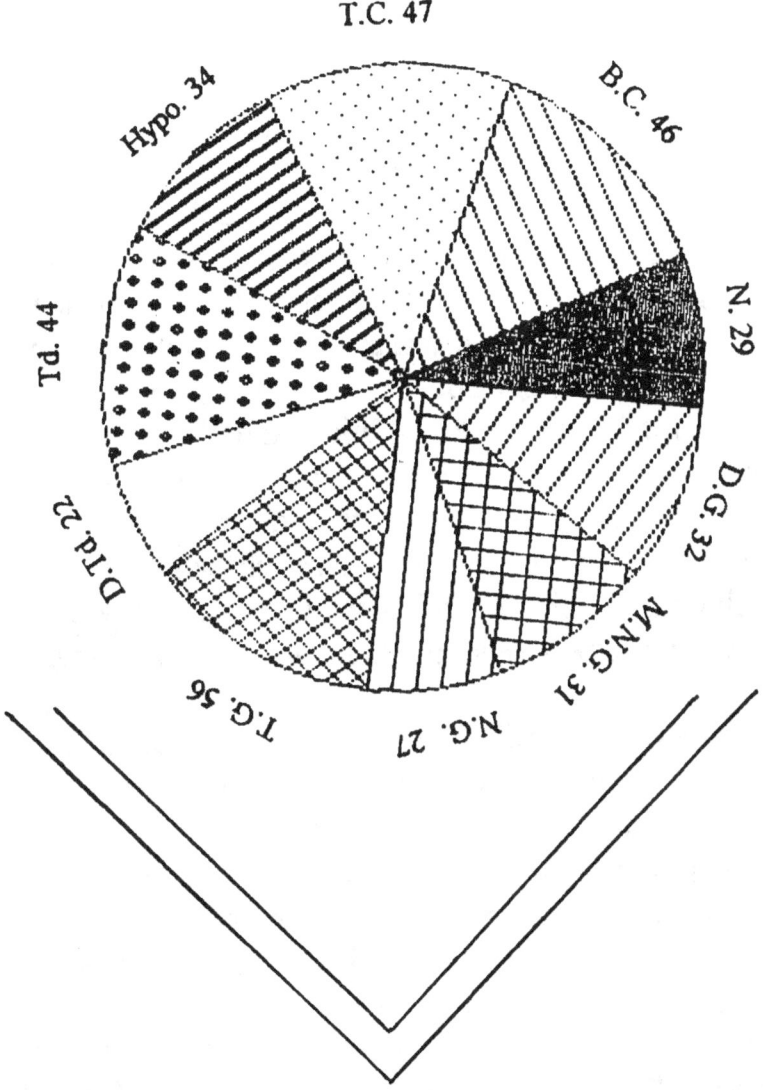

Fig. (3-4) : Distribution of LASA level in sera of normal volunteers (N),
pathological controls (BC), thyroid carcinoma, (TC),
hypothyroidism (Hypo), years after thyroidectomy (TD),
directly after thyroidectomy D.TD, toxic goiter (TG), nodular
goiter (NG) multinodular goiter (MNG), goiter (DG).

3-3   Determination of Seromucoid and serum protein bound hexose in sera of thyroid gland tumor diseases .

Figure (3-5) summarizes the distribution of the results obtained with respect to serum protein - bound hexose (galactose and mannose) in the sera of patients suffering from nodular goiter, multinodular goiter, toxic goiter, diffused goiter. hypothyroidism, several years after thyroidectomy, directly after thyroidectomy. thyroid carcinoma, and pathological control (Breast cancer), compared with normal. The average serum protein bound hexose in these diseases is significantly elevated Table (3-8) over that observed in normal healthy control.

Table (3-9) and Fig (3-6) summarizes the distribution and the results of seromucoid in the sera of the same patients used to determine protein - bound hexose. The average of seromucoid is significantly elevated over that observed in normal healthy contols. The elevation of seromucoid is higher in cases chosen after some years from thyroidectomy then followed by other pathological diseases. It is with no doubt that glycoproteins, secretions with excessive amounts in blood occur by major tissue alteration sites [138].

Table (3-10) and (3-11) show the specificity and sensitivity of seromucoid test depending upon 14 **mg/dl** as the upper limit of normal level . this type is least sensitive for thyroid cancer and pathological control (breast cancer).

Specificity and sensitivity of protein - bound hexose have been shown in table (3-12) and (3-13), dealing with 156 **mg/dl** as the upper limit of normal level . the test was least specific for thyroid carcinoma and pathological condition (breast cancer).

The increasing levels of protein - bound hexose arises due to depolymerization of the ground substance connective tissue adjacent to cancer with releasing these components in to blood circulation [174-179]. Hence the elevation of protein - bound hexose refers to depolymerization of the connective tissue adjacent to cancer with releasing these compound into blood stream.

Further more glycoprotein behaviors in cancer, non cancer disease and normal disease explain the protein metabolism abnormality, which is common to all tested group [180], and the investigation of chemical nature and physical properties of the isolated glycoprotein from pathological sera and compared with those of normal sera .

Seromucoid levels in table (3-9) indicate a significant elevation of (p value) in thyroid gland tumor disease and pathological control after comparison with normal healthy individual levels. table (3-8) also indicate elevation of significant value of total protein bound hexose. in hyperthyroidism, such as nodular goiter ($p < 0.01$). but in multinodular goiter. toxic goiter, thyroid cancer, sera patient underwent surgery and pathological control (breast cancer) with ($p < 0.005$) respectively increased while protein - bound hexose in hyperthyroidism: nodular goiter ($p < 0.01$). but multinodular goiter. toxic goiter. patients underwent surgery: thyroid carcinoma and pathological control ($p < 0.005$). within normal range. Estimation of glycoprotein level provide a clinical importance, the analysis of the result confirms previous observation about increasing of glycoprotein level in sera of thyroid gland tumor diseases with frequently modest elevation, so it seems to be unlikely useful diagnosis in case of thyroid carcinoma and breast cancer, this appear clearly by determination of protein - bound hexose. Moreover seromucoid estimation may be more specific than protein - bound hexose determination for thyroid gland tumor diseases and pathological control (breast cancer) when compared with normal healthy persons.

Table (3-8): Statistical evaluation of serum protein bound hexose level mg/dl ±SD in serum of thyroid gland patients :-

Group	No. of cases	Std. Deviation	Variant (Std.)^2	Average	Std. Error	Critical value of t-test	Significance value (P value).
Hypothyroidism:- Nodular goiter	13	2.61	6.8121	109.111	0.5836	2.7349	0.01
Multinodular goiter	14	2.348	5.5131	116.579	0.6275	6.220	→0.005
Toxic goiter	24	7.43	55.205	128.5	1.517	3.570	0.005
Goiter	12	3.879	15.047	107.797	1.1198	1.5014	0.05
Hypothyroidism	7	2.527	6.386	107.833	0.955	2.319	0.025
Years after thyroidectomy	5	3.485	12.145	120.2	1.559	5.230	0.005
Directly after Thyroidectomy	5	4.621	21.354	126.18	2.067	5.239	0.005
Thyroid carcinoma	10	5.74	8.237	137.0	0.908	4.252	0.005
Pathological contorl (Breast cancer)	10	8.241	67.914	134.5	2.600	3.947	0.005
Normal healthy persons	10	3.469	12.034	101.973	1.097	-	-

Table (3-9) : Statistical evaluation of seromucoid level mg/dl , ±SD in serum of thyroid gland patients :-

Group	No. of cases	Std. Deviation	Variant (Std.)^2	Average	Std. Error	Critical value of t-test	Significance value (P value).
Hypothyroidism:- Nodular goiter	13	0.454	0.6738	15.04	0.126	2.846	0.01
Multinodular goiter	14	0.726	0.5271	18.51	0.194	6.559	0.005
Toxic goiter	24	1.685	2.839	19.773	0.343	6.025	0.005
Diffused goiter	12	1.836	3.371	17.845	0.53	2.232	0.025
Hypothyroidism	7	1.026	1.053	17.933	0.388	4.079	0.005
Years after thyroidectomy	5	0.545	0.297	23.4	0.244	17.71	0.005
Directly after thyroidectomy	5	0.294	0.086	20.468	0.132	22.86	0.005
Thyroid carcinoma	10	2.87	8.237	20.912	0.908	2.496	0.01
Pathological control (Breast cancer)	10	1.42	2.016	24.7	0.449	7.713	0.005
Normal control	10	1.041	1.084	13.748	0.329	-	-

Table (3-10) : Specificity of seromucoid calculation were carried out by taking the number of cases which have 14% mg/100 ml or below levels then divided by total number of cases :-

Seromucoid Specificity	No. of cases	T≤14% mg/100ml normal	specificity true negative
Hyperthyroidism:- Nodular goiter	(13)	8	0.6
Multinodular goiter	(14)	4	0.28
Toxic goiter	(24)	6	0.25
Diffused goiter	(12)	4	0.33
Hypothyroidism	(7)	1	0.14
Years after thyroidectomy	(5)	0	-
Directly after thyroidectomy	(5)	5	1.0
thyroid carcinoma	(10)	3	0.3
Pathological contol	(10)	0	-
Normal level	(10)	8	0.8

Table (3-11) : Sensitivity of seromucoid calculation were carried out by taking cases which have ≥ 14% mg/100 ml value and divided by total cases :-

Seromucoid test Sensitivity	No.of cases	T≥14%mg/100ml	Sensitivity true positive
Hyperthyroidism , Nodular goiter	13	7	0.538
Multi nodular goiter	14	10	0.714
Toxic goiter	24	19	0.79
Diffused goiter	12	8	0.66
Hypothyroidism	7	6	0.857
Years after thyroidectomy	5	5	1.0
Directly after thyroidectomy	5	0	0
Thyroid carcinoma	10	7	0.7
Pathological control	10	10	1.0
Normal control	10	2	0.2

Table (3-12): Specificity of serum protein - bound hexose calculation were carried out by taking the number of cases which have 156% mg/100 ml or below levels, then divided by total number of cases:-

Protein – bound hexose specificity	No. of cases	T≤ 156% mg/100 ml normal	Specificity true negative
Hyperthyroidism, Nodular goiter	13	9	0.69
Multinodular goiter	14	3	0.93
Diffused goiter	12	11	0.916
Toxic goiter	24	15	0.6
Hypothyroidism	7	7	1.0
Years after thyroidectomy	5	5	1.0
Directly after thyroidectomy	5	5	1.0
Thyroid carcinoma	10	2	0.2
Pathological control	10	1	0.1
Normal control	10	9	0.9

Table (3-13): Sensitivity of serum protein - bound hexose calculation were carried out by taking the number of cases which have 156% mg/100 ml or elevated levels , then divided by total number of cases :-

Protein - bound hexose Sensitivity	No. of cases	T≥ 156% mg/100 ml elevated	Sensitivity true positive
Nodular goiter	13	4	0.308
Multinodular goiter	14	2	0.143
Diffused goiter	12	1	0.083
Toxic goiter	24	5	0.21
Hypothyroidism	7	0	-
Years after thyroidectomy	5	0	-
Directly after thyroidectomy	5	0	-
Thyroid carcinoma	10	8	0.8
Pathological control	10	9	0.9
Normal control	10	1	0.1

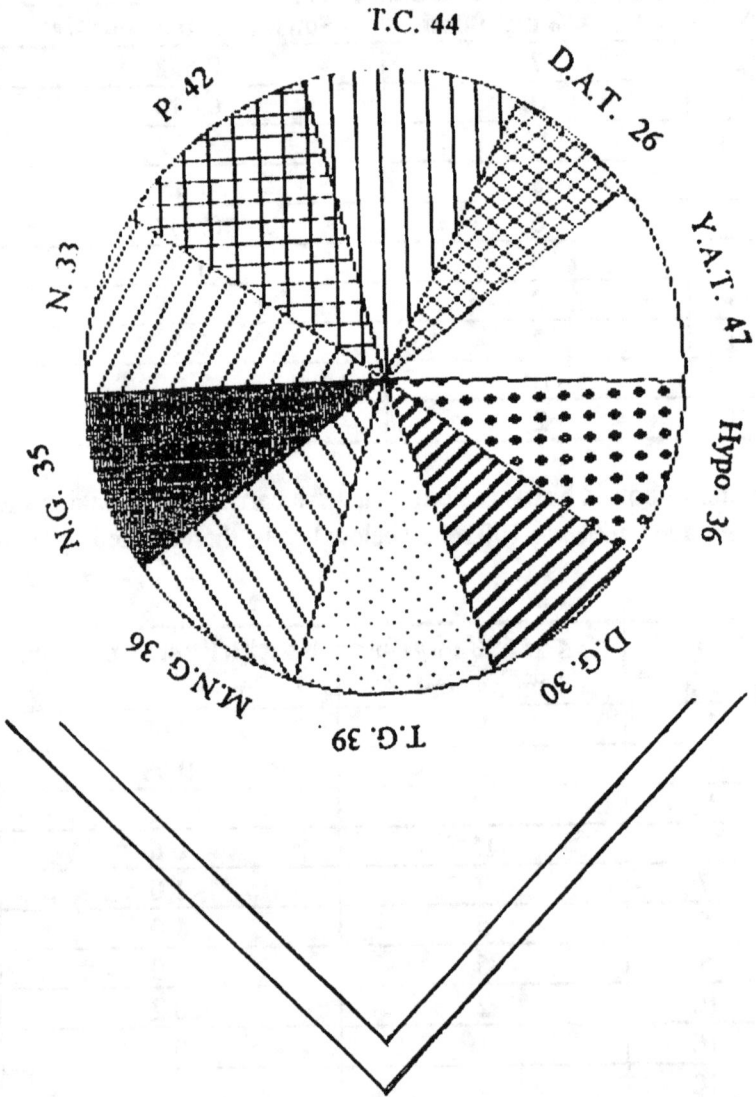

Fig. (3-5) : Distribution of serum protein bound hexose in sera of various thyroid gland diseases ; hyperthyroidism such as nodular goiter (NG), multinodular goiter (MNG), toxic goiter (T.G), goiter (D.G.), hypothyroidism (Hypo.), years after thyroidectomy (Y.A.T.), directly after thyroidectomy (D.A.T.) and thyroid carcinoma (T.C.) , compared with pathological controls (P) and normal persons (N).

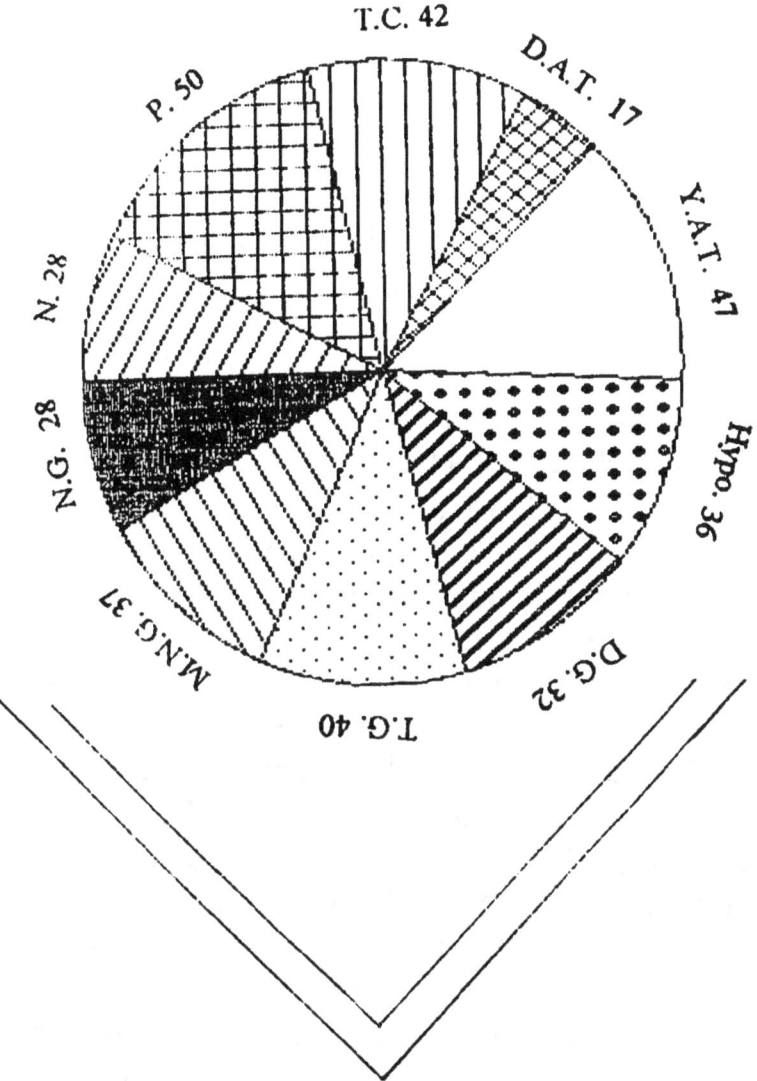

Fig. (3-6) : Distrbution of seromucoid in various thyroid gland diseases hypothyroidism (Hypo.), Hyperthyroidism such as nodular goiter (NG), multinodular goiter (MNG), toxic goiter (TG), goiter (D.G), years after thyroidectomy (YAT.), directly after thyroidectomy and thyroid carcinoma (TC) compared with pathological controls (P) and normal healthy persons (N).

3-4   Determination of hemagglutination activity for thyroid gland tumor lectin to red cells glycoprotein.

The extract of human thyroid tumor homogenate had a significant hemagglutination activities toward intact human erythrocyte group A. Glycoconjugate of thyroid follicular cell surface play important role in regulating the endocrine activity of this tissue [181]. the thyroid plasma membrane have been found to have three carbohydrate rich glycoprotein that show interaction with lectins [21,182-184]. these glycoproteins found to contain liked chains consists in both peripheral and internal location and contain sialic acid as well as D-galactosyl residues [185-186].

This investigation provide information in regard to the nature of crude thyroid tumors lectin binding with erythrocyte cell surface glycoprotein. These findings include the sampling that should be done at 4°C to avoid denaturation of proteins, and to prevent proteolytic enzymes activity [187], and to show the ability of lectin to agglutination [188-189]. Addition of 2-mercaptoethanol to achieve the maximal extraction of the lectin and to prevent inactivation of the lectin [190]. The active site of hemagglutination may be refers to N-acetyl group of siloglycoprotein of the erythrocyte outer cell surface, this occurrence has been observed in thyroid cell surface, that contain sialic acid [186]. As a result the thyroid cell may be enlarges the surface distribution which have been shown on erythrocyte cells binding thyroid gland tumor lectin. This suggestion is accepted in return to sialic acid residue which present on the erythrocytes surface that gives striking correlation between the ability of lectin to agglutinate cells and the presence of sialic acid residues on mammalian cell surface [117,186]. Figure (3-14) shows a rapid steep decline in hemagglutination activity % an infection point appeared were due to the thyroid lectin concentration used [191].

3-4-1   Optimum conditions of hemagglutination :-

The results obtained show that lectins of thyroid gland tumors are temperature sensitive, undergo agglutination of 25°C and at pH 9 for 16 minutes figures (3-7, 3-8, 3-9). The low hemagglutination was observed at pH 7 Figure (3-7) due to sialic acid of red cell outer membrane which becomes unstable at this pH [186] and alkaline pH.

Generaly the binding process is pH dependent and pH effect includes the induction of protonation - deprotonation process occurring within the ionizable groups of the amino acids present in the binding groups of the lectin binding complex [192].

3-4-2 The effect of various carbohydrates on hemagglutination activity:-

Thyroid tumor lectin is a potentially useful agent for *in vivo* assessment of thyroid gland disorders. The multivalent ligand cause precipitate with lectins non - specific interaction by sugar inhibitions [193-194]. Glucouronic acid was used at high level determinate for non - specific binding of thyroid gland tumor with RBC (figure 3-10) [195].

A number of carbohydrates were tested for hemagglutination activity toward thyroid gland tumor lectin homogenate Fig. (3-14) using D-fructose , D-galactose, D-xylose, D-mannose and sucrose at 30mM content . D-fructose, D-galactose, D-xylose and sucrose slightly depressed the hemagglutination activity, while D-mannose has no effect on the hemagglutination activity and at high concentrations will cause inhibition of hemagglutination activity.

3-4-3 The role of δ-globulin on hemagglutination activity:-

Hemagglutination activity of crude thyroid tumor lectin to red cell glycoprotein is inhibited by δ-globulin immunoglycoprotein containing sialic acid residue Fig (3-11), show the high affinity of thyroid gland multinodular goiter tumor lectin to the sialyliated glycopeptides of the immunoglycoprotein at 2.5 ng/ml of δ-globulin concentration .

3-4-4 Effect of Ca^{++} ion on hemagglutination activity :-

Like other sialic acid - binding Lectin [195], this lectin required divalent calcium ions for binding activity or hemagglutination . It is known that the Ca^{++} is enhansing the binding activity of lectin [195]. Our results have shown that the level of Ca^{++} for multinodular goiter was higher than nodular goiter table (3-15).

The conformational alternation in lectin binding sialoglycoprotein *in vivo* due to calcium ion binding which produce a stable complex [196]. The presence of calcium ion in thyroid gland maintain lectin in its most highly aggregated form. Figure (3-12) indicate no optimum hemagglutination activity observed at increasing $CaCl_2$ concentration that was similar to the results obtained George Lee *et al* [197] study and Gambino *et al* 1997 [198]. The other divalent salts $MnCl_2$, $MgCl_2$ have shown a slight increase in the hemagglutination activity figures (3-13 : c,d).

While the monovalent ions of NaCl & KCl indicated in figures (3-13: a,b) have shown a decreased in the hemagglutination activity.

Figures (3-16: a-b) represent the hemagglutination activity change on adding different concentrations of urea polyethylene glycol. The lectin was unfolded by disulfide bond fission and for urea and conformational alteration due to polyethylen glycol effects.

Table (3-14): Effect of various carbohydrates on hemagglutination activity of thyroid gland Lectin homogenate to red cells glycoprotein

Type of carbohydrate 30 mM	Hemagglutination activity %
Control	90.
D-Frcuctose	82.85
D-Xylose	83.1
D-galactose	83.25
Sucrose	83.3
mannose	89.75

Table (3-15): Calcium ion concentration in thyroid gland tumor homogenate by as determined by Atomic absorption spectroscopy.

Type	Calcium ion concentration p.p.m.	Calcium ion concentration mg/L
Multinodular goiter	1.525	1.525×10^{-3}
Nodular goiter	0.9	0.9×10^{-3}

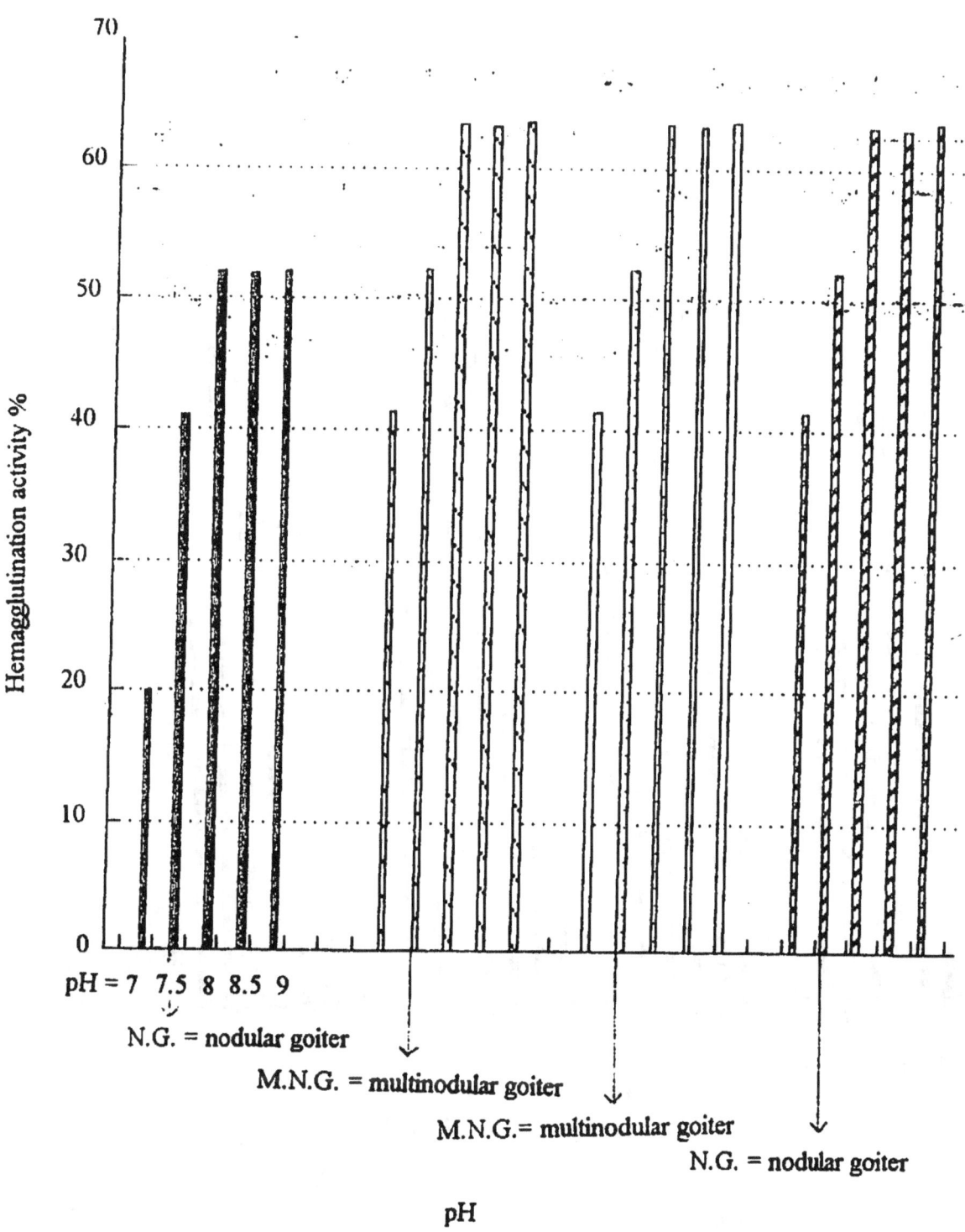

Fig. (3-7) : Effect of various pH on different thyroid gland lectin binding
           complex. (All details explained in 2.12.5).

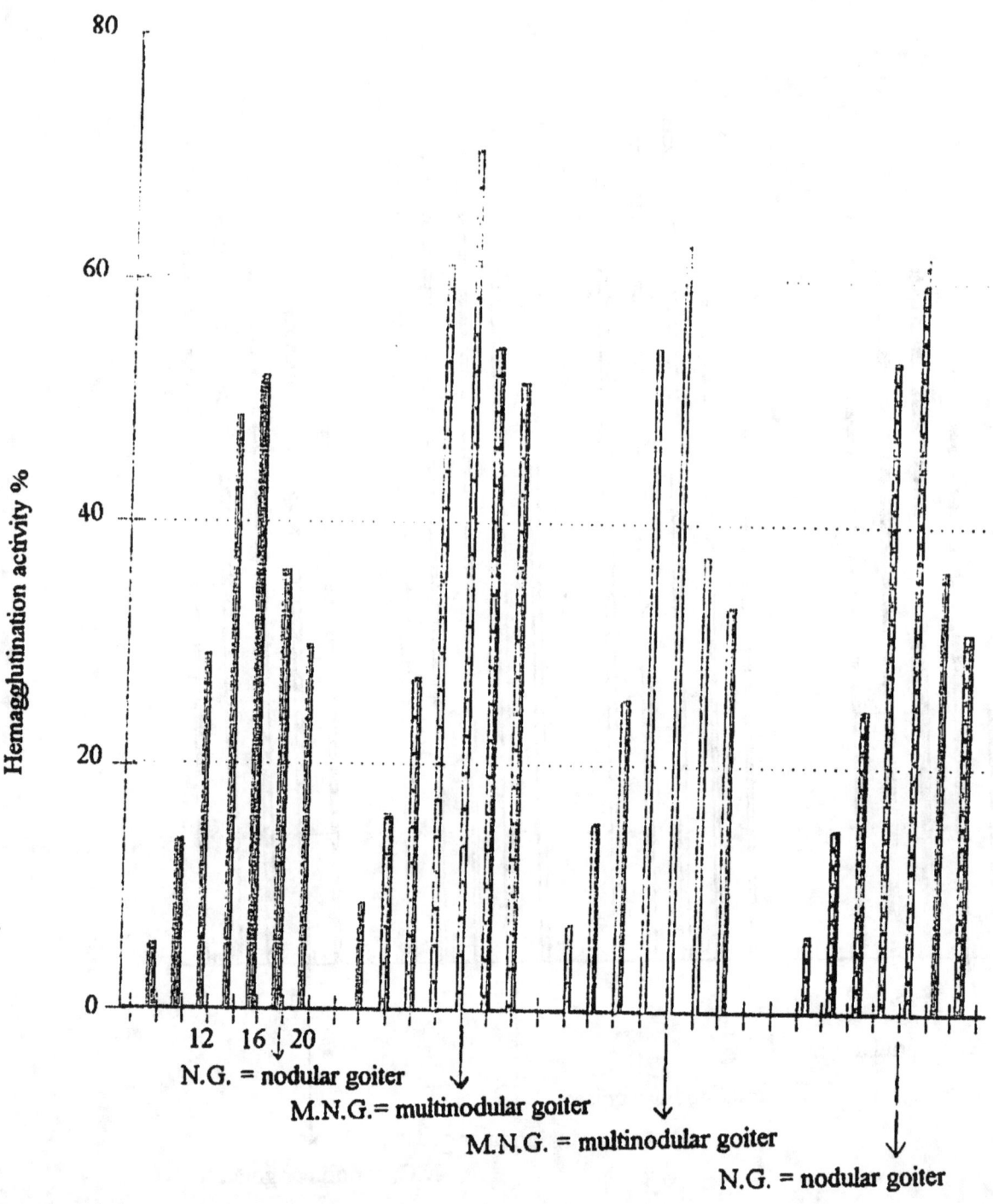

Fig. (3-8) : Dependence of hemagglutination activity on time for nodular goiter (NG), and multinodular goiter (MNG), lectin homogenate. (All details explained in 2.12.7).

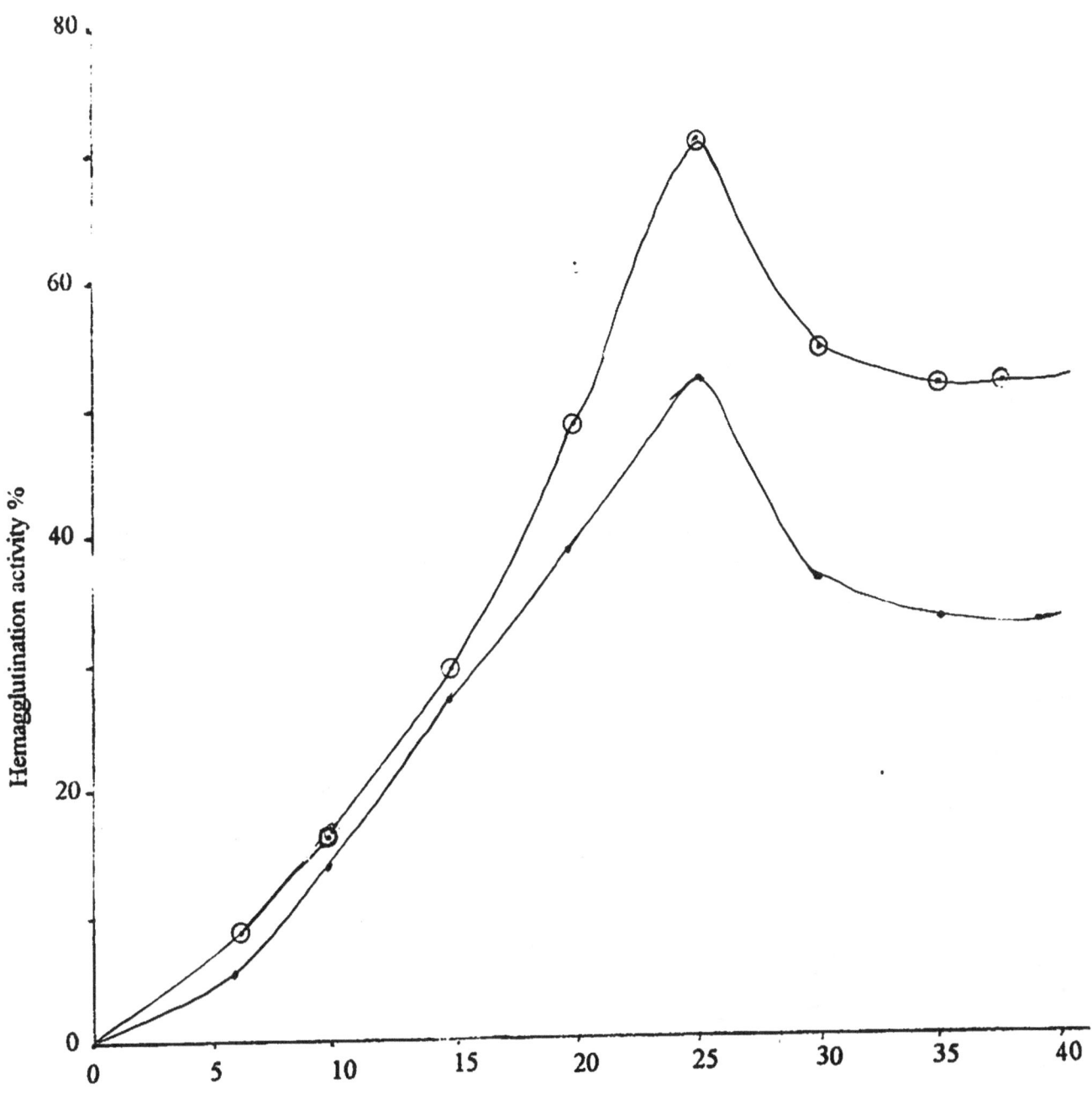

Fig. (3-9) : Dependence of hemagglutination activity of human thyroid
            tumors   nodular goiter (•NG) and multinodular goiter
            (⊙MNG) lectin on temperature. (All details explained
            in 2.12.6).

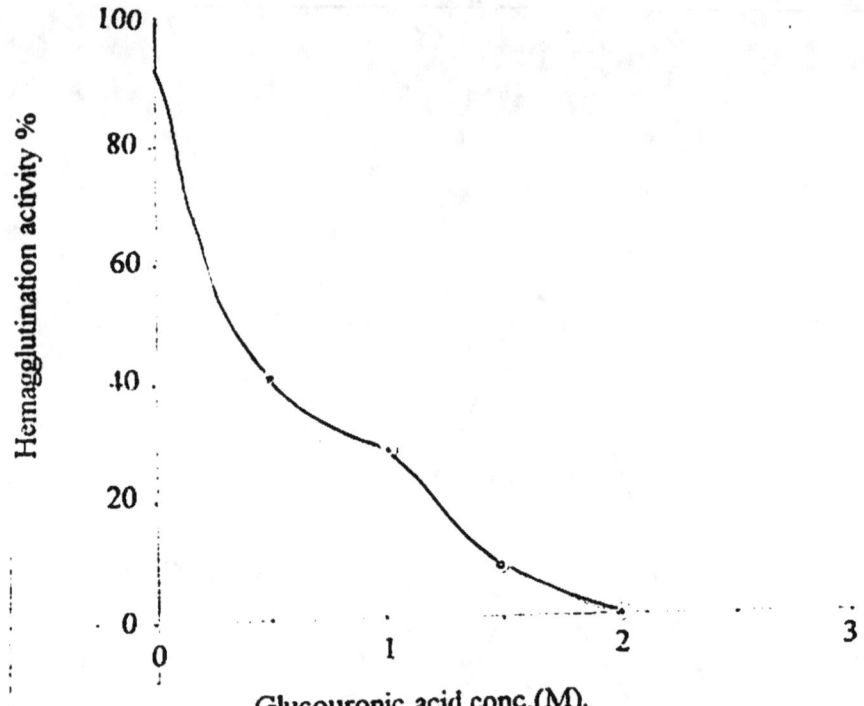

Fig. (3-10) : Determination of non specific binding of thyroid tumor lectin binding complex by glucouronic acid. (All details explained in 2.12.3).

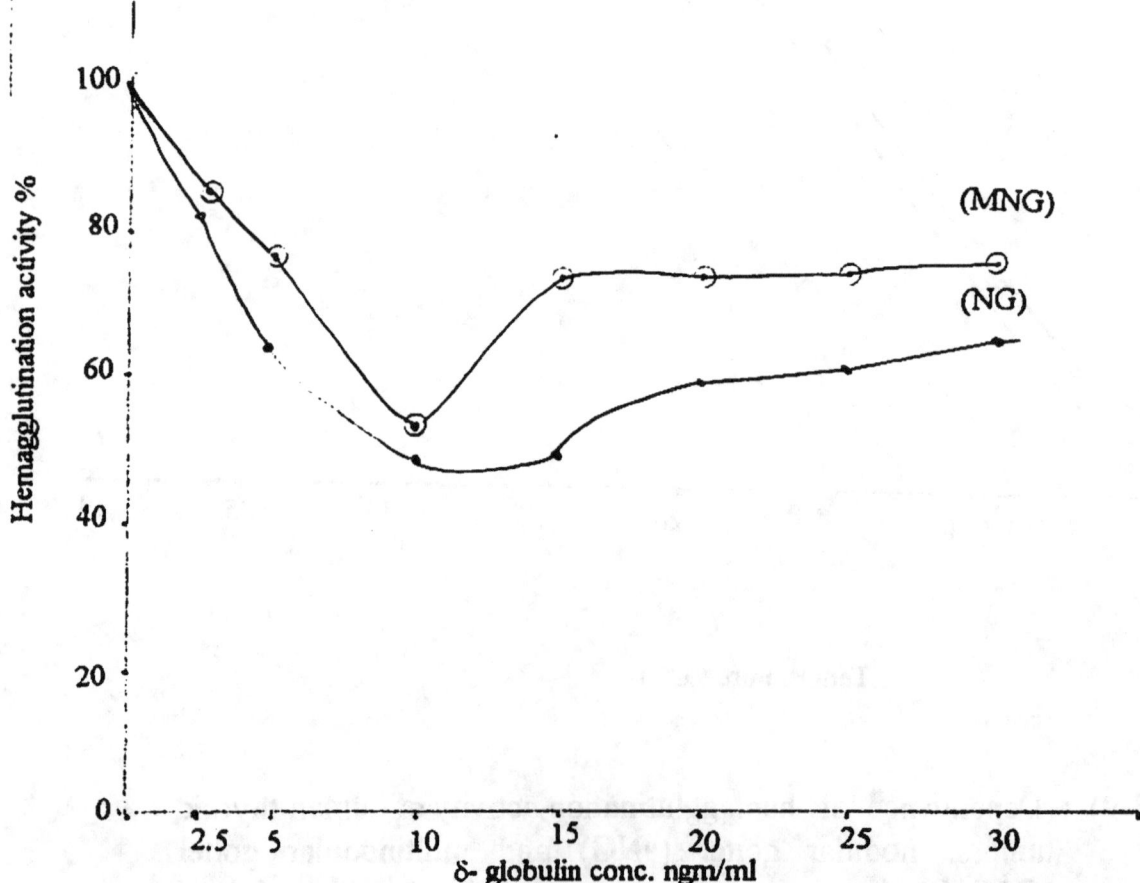

Fig. (3-11) : Effect of different δ- globulin concentrations on hemagglutination activity of thyroid gland tumor(•NG) nodular goiter & (⊙MNG) multinodular goiter lectin. (All details explained in 2.12.14).

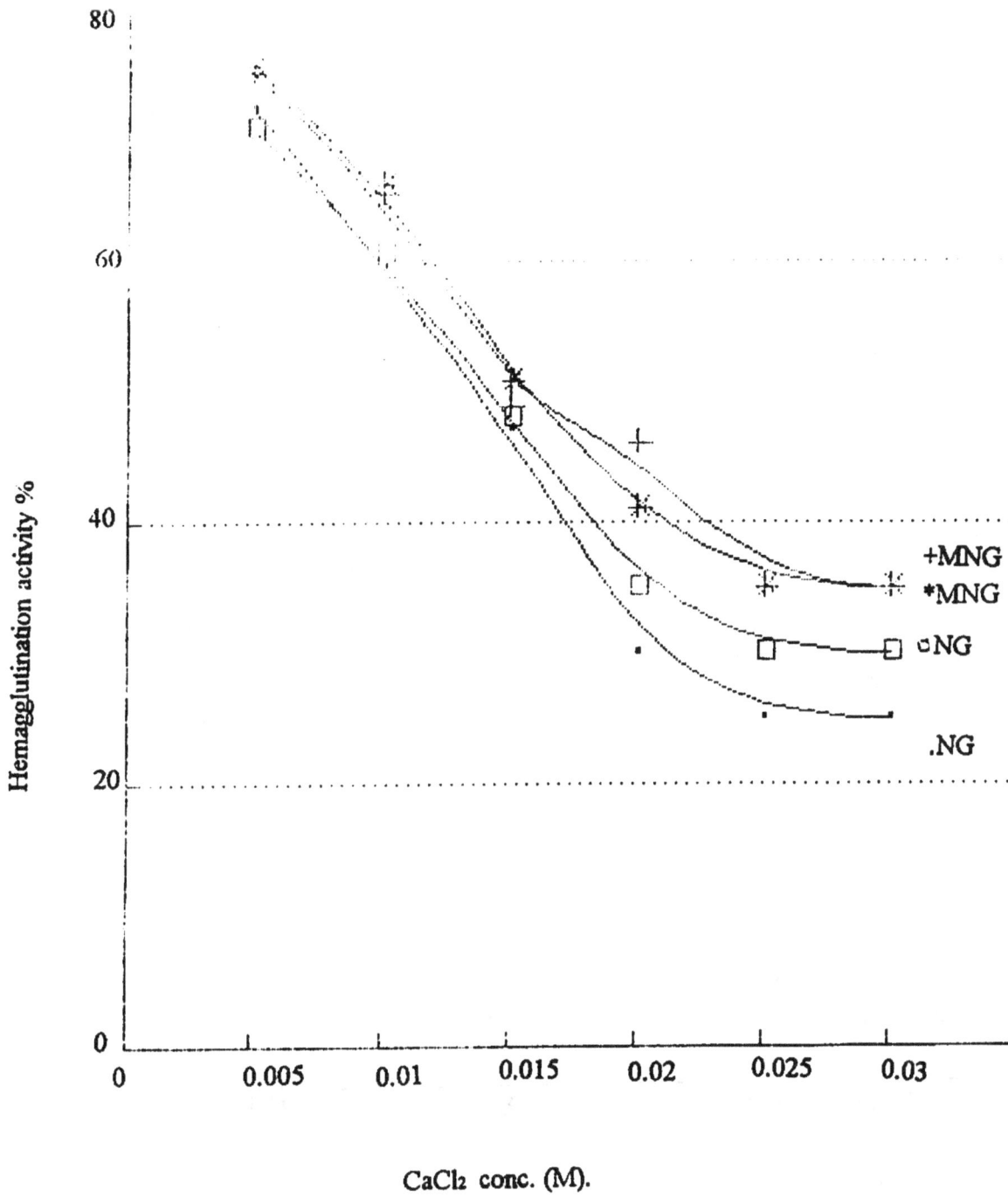

CaCl₂ conc. (M).

Fig.(3-12): Effect of increasing CaCl₂ concentrations on hemagglutination
activity of different thyroid tumors lectin homogenate.
(All details explained in 2.12.9).

□N.G. = nodular goiter ⊹M.N.G.= multinodular goiter ⋇M.N.G. = multinodular goiter
•N.G. = nodular goiter

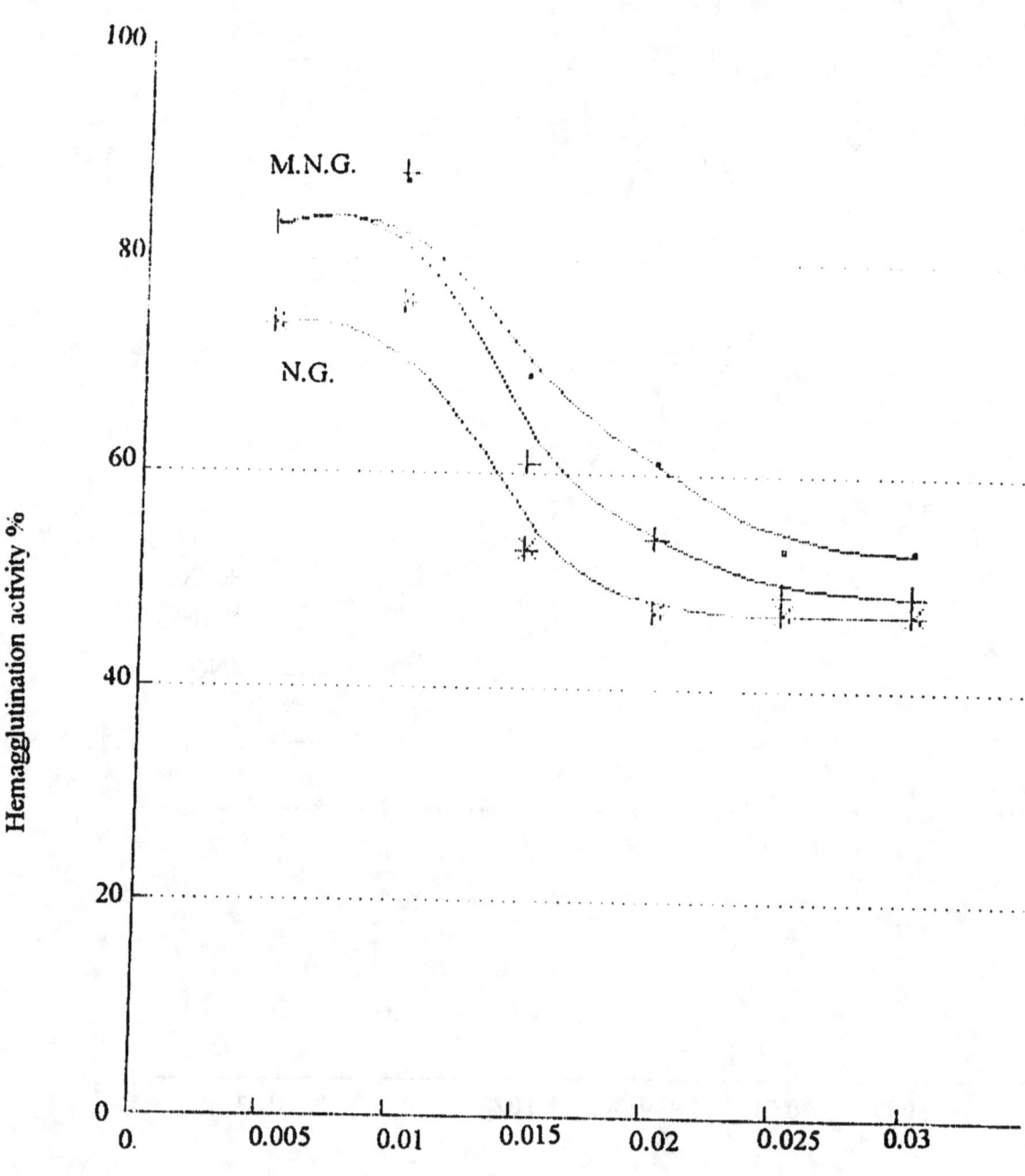

Fig. (3-13a): Effect of monovalent salt (NaCl), on hemagglutination activity of NG and MNG thyroid tumor lectin binding complex. (All details explained in 2.12.11).

N.G. = nodular goiter   M.N.G.= multinodular goiter

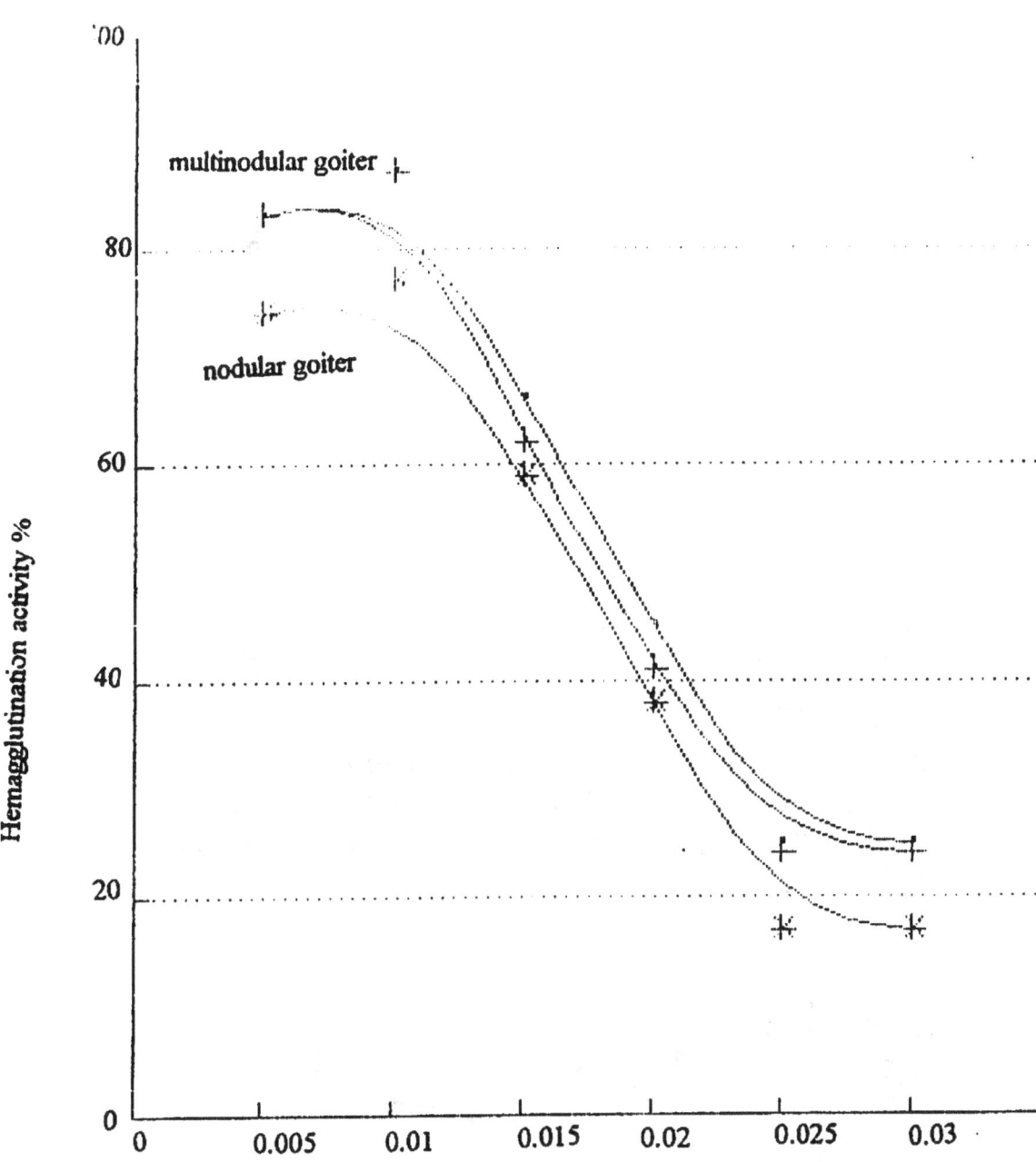

KC l conc. (M).

Fig. (3.13b) : Effect of monovalent salt (KCl), on hemagglutination activity of NG, and MNG thyroid tumor lectin binding complex .(All details explained in 2.12.11).

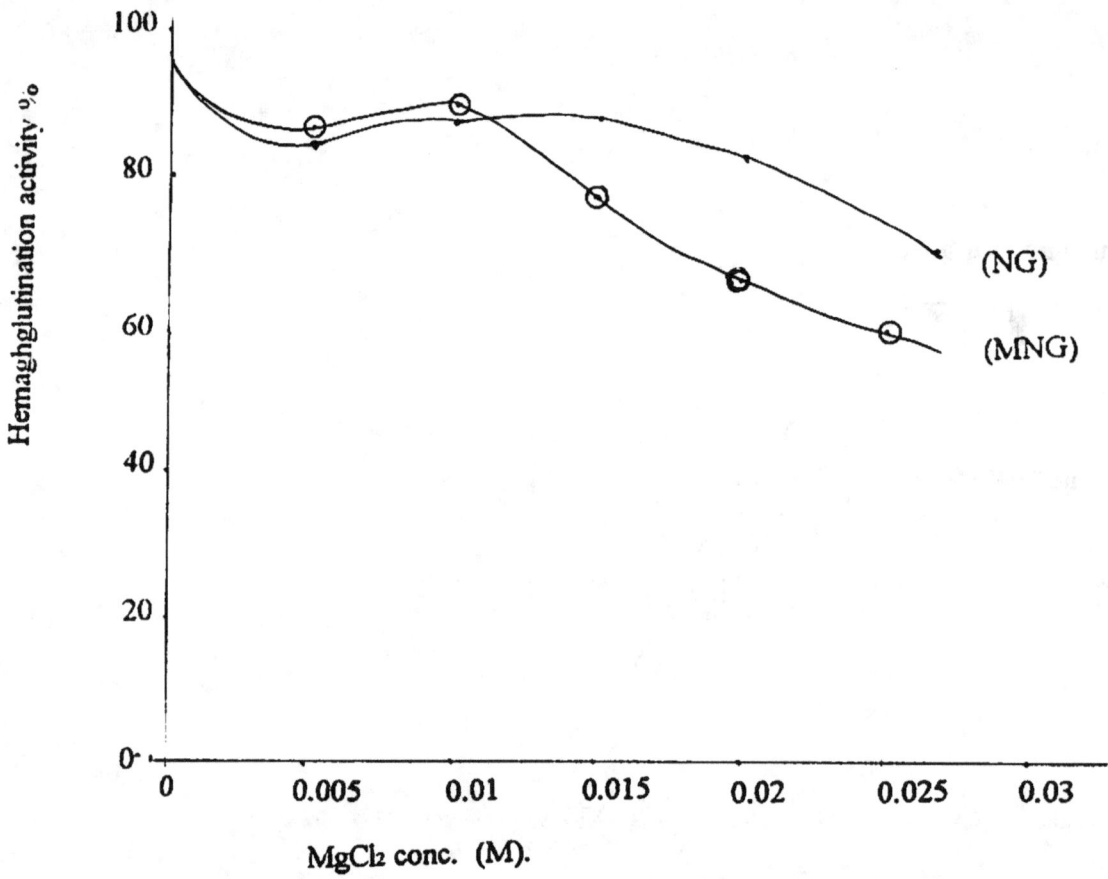

Fig. (3-13c) : Effect of MgCl₂ divalent salt on hemagglutination activity of thyroid gland tumor(•NG) nodular goiter & (⊙MNG) multinodular goiter lectin. (All details explained in 2.12.12).

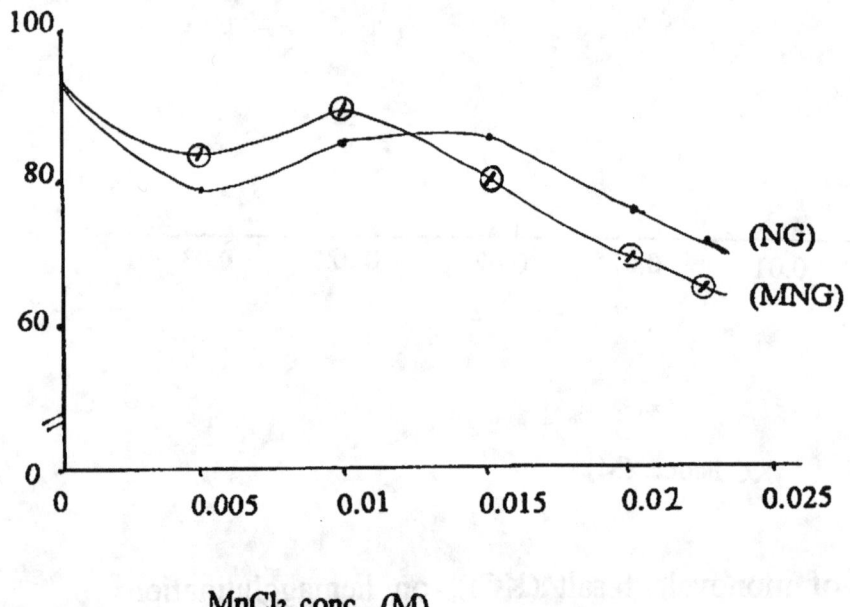

Fig. (3-13d) : Effect of MnCl₂ divalent salt on hemagglutination activity of thyroid gland tumor(•NG) nodular goiter & (⊙MNG) multinodular goiter lectin. (All details explained in 2.12.12).

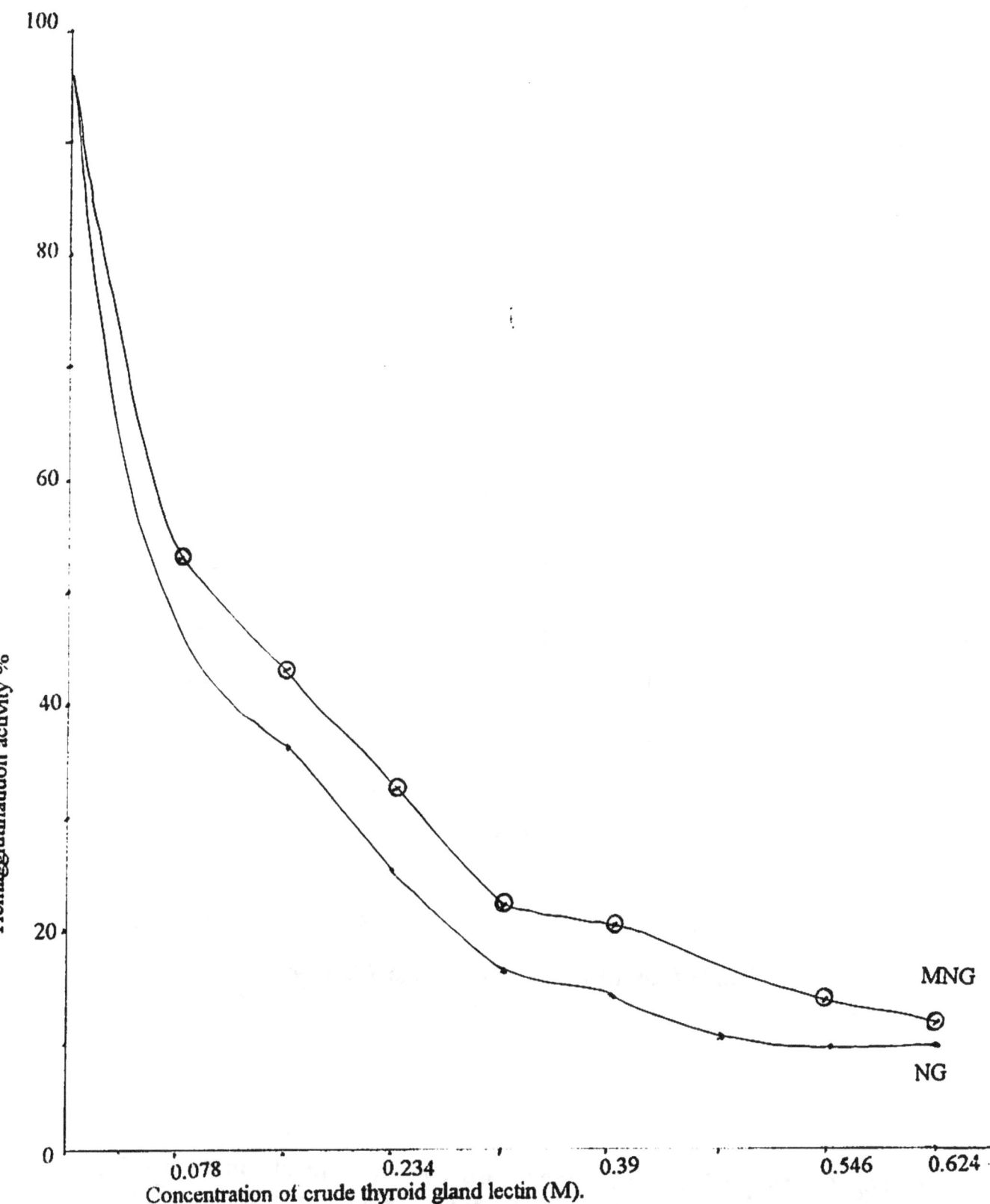

Fig. (3-14) : Hemagglutination activity assay of thyroid gland tumor
lectins of nodular goiter •(NG) and multinodular goiter
⊙(MNG), the crude approximately (78 μg 0) protein gave a
50% in hemagglutination unit (HU).

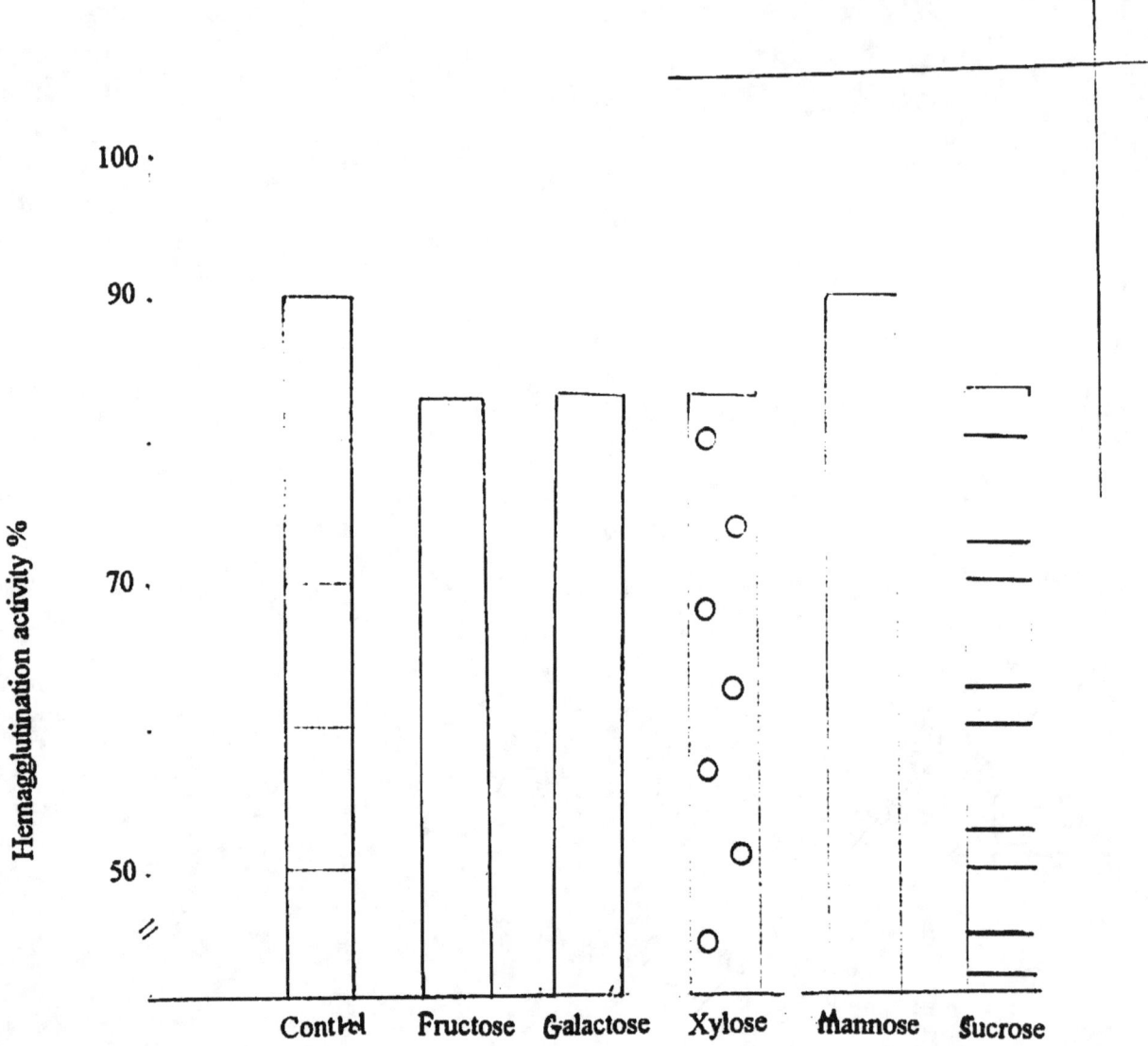

Kind of carbohydrate used at conc. (30mM).

Fig. (3-15) : Effect of various carbohydrate with 30 mM conc. on
hemagglutination of crude thyroid gland lectin (MNG)
(All details explained in 2.12.13).

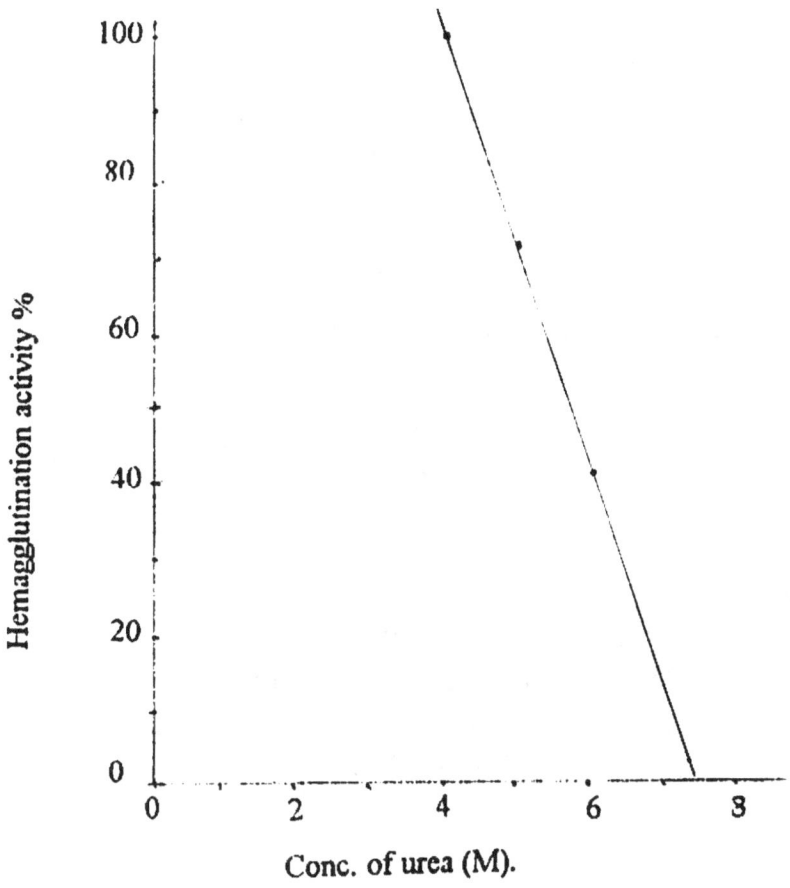

Fig (3-16a): Effect of denaturating agent (urea) on hemagglutination activity. (All details explained in 2.13 10).

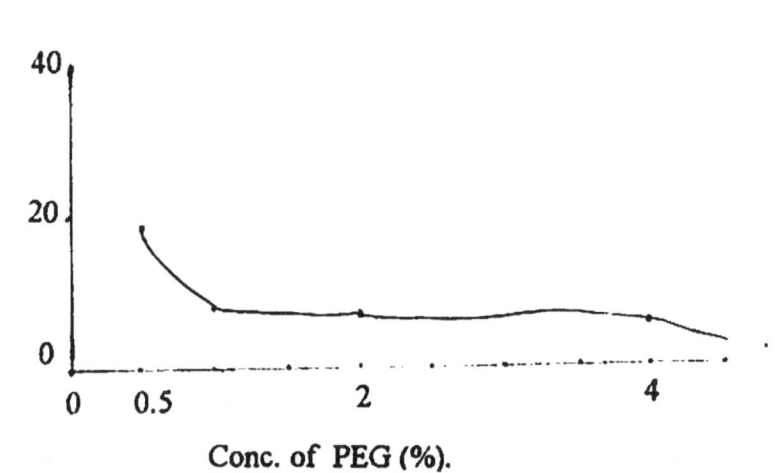

Fig.(3-16b): Effect of denaturating agent (PEG) on hemagglutination activity. (All details explained in 2.13 10).

3-5 Purification and identification of human thyroid gland tumor lectin:-

Purification of human thyroid gland tumor lectin was performed by adsorption to formalinzed human erythrocytes. then elution with D-glucouronic acid and subsequent fractionation on sephadex G-150. Tabel (3-16) Figure (3-17) describe calibration proteins from sephadex G-150. The elution volume Ve and K_{av} value. for thyroid   gland tumor lectin homogenate from the gel was calculated and found to be 53 ml and 0.8148 respectively.

The application of K_{av}. value to the calibration curve Figure (3-19) indicate that the stock radius (Rs) were 28.8 A° and the molecular weight MW = 40180 Dalton.

From the obtained results of SDS - gel electrophoresis. there is one lectin in this thyroid gland tumor (nodular goiter for example) as in Figure (3-18 a). The molecular weight was measured by two completely different methods, gel filteration and electrophoresis techniques.

In Figure (3-18 b) represent the calibration curve for SDS PAGE 7.5% using law molecular weight groups of known subunets proteins in the presence of 1% SDS.

The application of R_m value for each standard protein to calibration curve, the molecular weight of lectin was estimated and found to be 44157.05 Dalton. The purified glycoprotein possessed low electrophoretic mobility. Fig. (3-19).

Table (3-16): Purification of lectin from thyroid gland tumor. The hemagglutination
activity obtained by each step was determined as described under method.

Fraction	Protein mg/ml	Total activity unit X1000	Purification	%Recovery
Crude Cell extraction of thyroid tumor	23.3	2000	1	100
Elute from Red Cells	0.1623	1731	26.33	86.56
concentration of elute	0.2297	204.	224.69	12.2
Pooled peack from sephdex column	0.468	1248	29.53	62.4

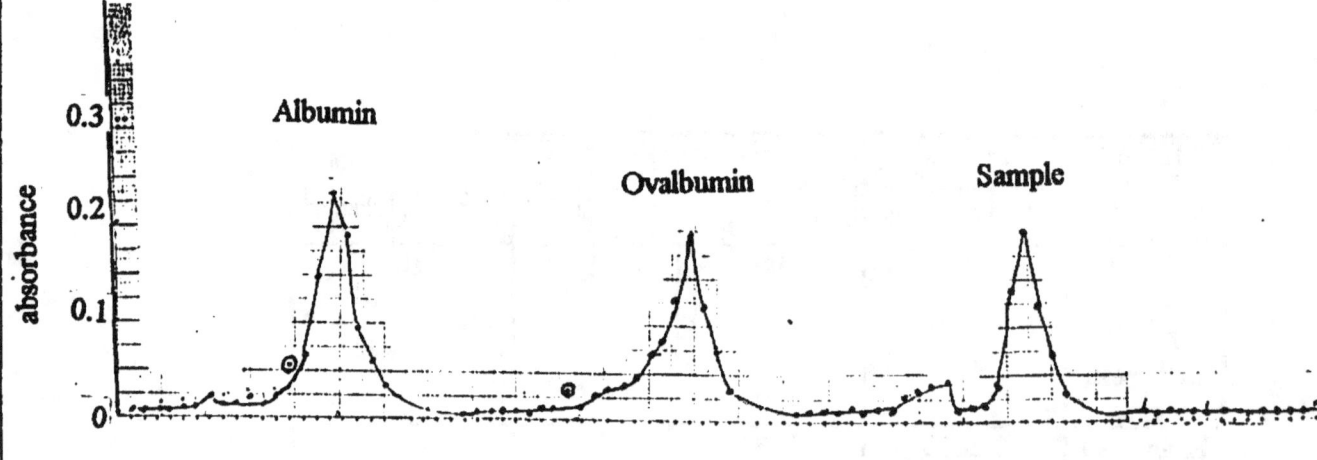

Fraction no. each fraction = 3ml.

Fig. (3-17) : The elution profiles of standard protein and thyroid gland
tumor lectin from sephadex G-150 column (1.5 X 7.5 cm
All details are outlined in the text .

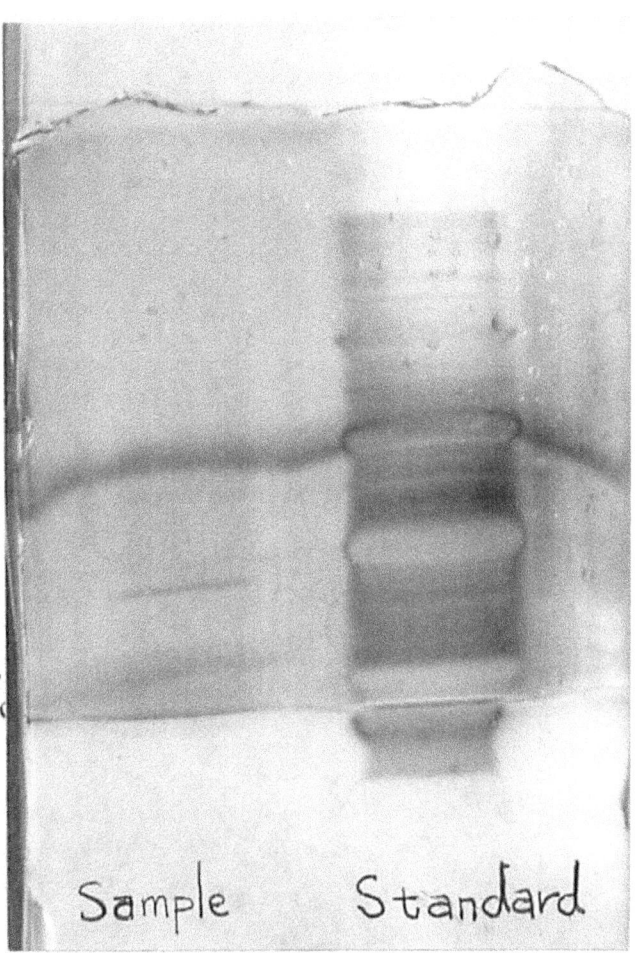

Fig. (3-18a) : The elecrophorsis migration of purified thyroid gland tumor lectin Right represent marker standard solution bands, the first refer to phosphorylase b, then BSA, ovalbumin,carbonic anhydrase Trypsin inhibitor and α- Lactoalbumin.

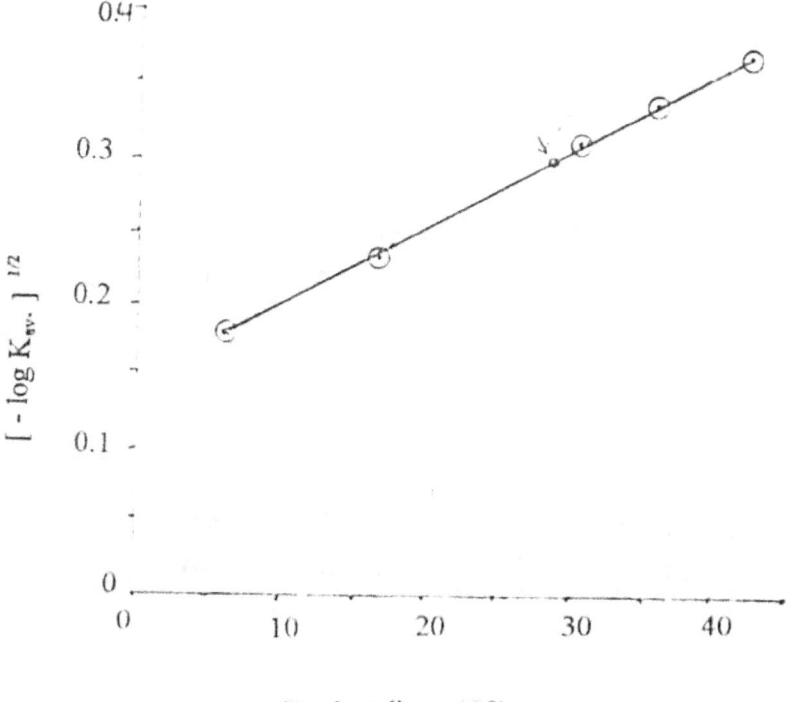

Stock radius (A°)

Fig. (3-19) : Determination of stock radius of purified thyroid tumor lectin.

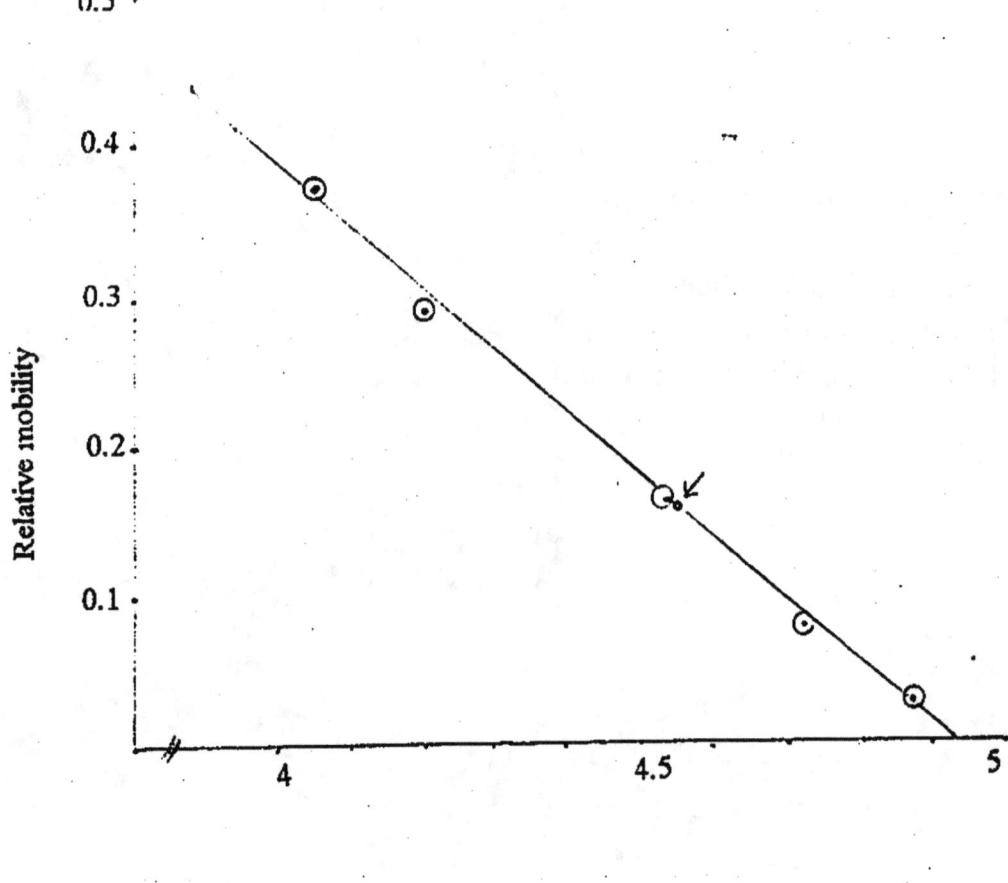

Fig. (3-18b) : Estimation of purified thyroid tumor lectin molecular
weight by electrophoresis mobility using marker standard
proteins.

3 6 The kinetics and thermodynamic studies of thyroid gland diseases Lectin.

1- The kinetics of the binding of human thyroid gland tumor lectin to red cell surface glycoprotein:-

Figures (3-20: a-b) show the time course of the formation of thyroid gland tumor lectin - glycoprotein complex at 25°C for both nodular goiter and multinodular goiter. The concentration of lectin - glycoprotein complex formed was calculated after time (t) by the following:

$$\text{Concentration of Lectin-glycoprotein formed after (t) time in M} = \frac{A_{620} \text{ (red cells+Lectin)}}{A_{620} \text{ (red cells)}} \times A_{620} \text{ (red cells+Lectin)} \times \text{conc. of Lectin in M}$$

The results obtained of the time course indicate the dependence of the binding on the time and temperature (25°C).

Determination of kinetic parameters of lectin-glycoprotein complex formation :-

The time course of lectin - glycoprotein complex formation depends on the binding kinetic parameters[164] (expressed as hemagglutination). The following reaction indicate the simplest proposed model of the complex interaction: -

$$\text{Lectin} + \text{Glycoprotein} \underset{K_{-1}}{\overset{K_{+1}}{\Longleftrightarrow}} [\text{Lectin - glycoprotein}] \text{ complex} \dots\dots (1)$$

(thyroid gland tumor)   (red cells)

K_{+1} : is the association rate of lectin to glycoprotein .
K_{-1} : is the dissociation rate of this complex at equiblibrium .

$$K_a = \frac{[\text{Lectin - glycoprotein}]}{[\text{Lectin}][\text{Glycoprotien}]} \dots\dots\dots\dots (2)$$

$$K_d = \frac{[\text{Lectin}][\text{Glycoprotein}]}{[\text{Lectin - glycoprotein}]} \dots\dots\dots\dots (3)$$

Thus : $Ka = \dfrac{1}{Kd} = \dfrac{K_{+1}}{K_{-1}}$ .... (4)

Where

    Ka : is the equilibrium constant (affinity constant)
    Kd : is the equilibrium constant of complex dissociation .

The total concentration of lectin binding sites (B_{max}), K_a and K_d values were calculated by scattered plot Figure (3-21) and Line weaver – Burke reciprocal plot Figure (3-22), at 25°C and at the time of incubation.

Figure (3-20 b) represent the time course data, which fit the first order kinetics for the association of lectin - glycoprotein, but due to the biomolecularity of this reaction, equation (5) could be simplified to equation (6) in order to fit the data of the first order kinetics :-

$$\ln [\text{Lectin - glycoprotein}]_{eq}\left[\frac{(\text{Lectin})_T - (\text{Lectin-glycoprotein})_t \cdot (\text{glycoprotein})_T}{(\text{Lectin})_T [\text{Lectin-glycoprotein})_{eq} - (\text{Lectin -glycoprotein})_t}\right]$$

$$= K_{+1} \ t \ \frac{[(\text{Lectin})_T (\text{glycoprotein})_T - (\text{Lectin - glycoprotein})_{eq}]}{(\text{Lectin - glycoprotein})_{eq}} \quad .... (5)$$

$$\ln\left[\frac{(\text{lectin - glycoprotein})eq}{(\text{lectin-glycoprotein})eq - (\text{Lectin-glycoprotein})t}\right] = K_{+1} \ t \ \frac{(\text{Lectin})_T (\text{glycoprotein})_T}{(\text{Lectin-glycoprotein})_{eq}} \ .... (6)$$

While:

[Lectin-glycoprotein]$_{eq}$: is the concentration of complex formation at equilibrium .

[Lectin-glycoprotein]$_t$: is the concentration of complex formation after time (t) .

K_{+1} : is the kinetic association constant in $M^{-1} min^{-1}$.

Equation (6) was obtained by simplification of equation (5) which corresponds to second order reaction kinetics, equation (6) was used for two iomportant reasons :-

Although most of lectin remained as a free form & only a small amount of lectin is bound even at equilibrium Pseudo - first order conditions of the binding

data i.e., $\ln \left(\dfrac{(\text{Lectin – glycoprotein})_{eq}}{(\text{Lectin - glycoprotein})_{eq} - (\text{Lectin - glycoprotein})_t}\right)$ vs. t, a straight line

was obtained with a slope equal to the observed constant value (K_{obs}) in min^{-1} (first rate constant).

The value of K_{+1} was calculated from the following equation :-

$$K_{obs} = K_{+1} \frac{[\text{ Lectin }]_T [\text{ glycoprotein.}]_T}{[\text{ Lectin - glycoprotein.}]_{eq}}$$

$$\therefore K_{obs} = K_{+1} [\text{ Lectin }]_t.$$

The value of $K_{obs} = 0.0873$ min^{-1} and $K_{+1} = 0.0453 \times 10^6$ M^{-1} . min^{-1} at 25°C . While the results obtained at 20°C for $K_{obs.} = 0.0809$ min^{-1} and $K_{+1} = 0.0419 \times 10^6$ M^{-1} min^{-1}. The results indicate the dependence of reaction rate on temperature Figures (3-20: a-b) and table (3-17) represent these findings.

3-6-1-1   Scatchard and line weaver - Burke Analysis

Figures (3-21) and (3-22) show Scatchard plot and Line weaver - Burke plot of thyroid gland lectin to glycoprotein at 25°C. From scatchard plot the values of Kd, Ka. obtained were Bmax $0,2857 \times 10^7$ M ,3.5×10^{-1} M^{-1} ,2.5×10^{-7} M respectively. The values of Bmax obtained from Line weaver -Burk plot have shown there dependence on temperature and indicate that the highest maximum capacity was at 25°C thyroid. This interaction may be due to increases the number of molecules processing the activation energy for binding with temperature Ka value are also depend on temperature, this may be refer to the endothermic reaction of the complex formation and the affinity reaction enhanced by increasing temperatures .

3-6-1-2   Determination of Hill coefficient (n) of lectin - glycoprotein interaction :-

When the results gained by line weaver - Bruk analysis ,were applied to the Hill equation [199-201] .The cooperatively of the binding sites of lectin was observed Figure (3-23) the value of n = 1.25, indicates that the number of binding sites receptor exhibiting weak positive cooperatively .

3-5-2 The thermodynamic of the binding of thyroid gland tumor to its glycoprotein :-

A- The thermodynamic parameters of standard state :-

Table (3-18) show the dependence of the equilibrium binding constant (i.e. affinity constant) of binding of thyroid gland lectin to glycoprotein on temperature ..$\Delta H°$ value was positive , which mean the binding reaction was nearly endothermic , concluding that the reaction spontaneity was entropically driven ; so the entropy was the driven force of the interaction occurrence .

The negative value of $\Delta G°$ reflects the stability of the complex or the high interaction of the binding reaction. The high negative $\Delta G°$ value of the affinity reactions are controlled by the positively $\Delta S°$ as table (3-18) show , so the reaction system is characterized by the contribution of $\Delta S°$ to the complex formation stability . While $\Delta H°$ indicate little effect, the positive $\Delta H°$ value reflect the favorable interaction between groups within both thyroid gland lectin and erythrocyte glycoprotein .

The negative value of $\Delta G°$ represent the overall affinity reaction of lectin - glycoprotein complex formation was energetically favorable.

B- The thermodynamic parameters of transition state:-

The transition state of thermodynamic parameters as in table (3-18) shows association rate dependence of the interaction of thyroid gland lectin to glycoprotein on temperature , fig.(3-24): (Arrhenius plot) .

The highly positive ΔG^* value represent the formation of an activated complex which was a non - spontaneous process and required a lot of energy = Ea to overcome the transition state energy to provide the final product complex , while the high negative ΔS^* value represent that the activated complex which had a more ordered structure than the reactants . The positive ΔG^* value is attributed to the decreases in entropy of the transition state .

Determination of binding - reaction thermodynamics using equilibrium data represent an overall idea about the nature of forces controlling complex formation [200,202,203] . The comparison of ΔG^* , ΔH^* and ΔS^* of the transition state with $\Delta G°$, $\Delta H°$ and $\Delta S°$ in table (3-18) lead to choose the thermodynamic model.

Figure (3-25) indicate the existence of three thermodynamic states, the initial state A represent the starting energy level of the isolated hydrated species, in the thermodynamic state B, the lectin and glycoprotein come together and mutually penetrated their hydration sphere to form partially immobilized hydrophobically associated species, and thermodynamic state (indicates the fully interacting lectin state (indicates the fully interacting glycoprotein complex).

Table (3-17): The effect of temperature on the kinetic parameters of the binding of binding thyroid gland lectin to glycoprotein :-

T°C	K_{obs} (min^{-1})	K_{+1} (min^{-1}) M^{-1}	K_{-1} (min^{-1})	$Ka = \frac{k_1}{k_{-1}}$ M^{-1}	$Kd = 1/Ka$ M
25	0.0873	0.00453X10^7	0.07788	0.05823X10^7	17.173X10^{-7}
20	0.0809	0.00419X10^7	0.0720	0.05817X10^7	17.191X10^{-7}

Table (3-18): Thermodynamic parameters for the binding of thyroid gland lectin to glycoprotein at transition state :-

T(k)	ΔG° (J.mole^{-1})	ΔH° J.mole^{-1}	ΔS^α J.mole^{-1} K^{-1}	Ea J.mole^{-1}
279	-3088.5			
288	-37293.1			
293	-37934.96	323.77	130.58	
298	-38725.42	8375.29	+158.	
	ΔG^* (J.mole^{-1})	ΔH^* J.mole^{-1}	ΔS^* J.mole^{-1} K^{-1}	
	44008	1009	-139.6	76.737

In the first step A the reaction binding was associated with positive ΔG^* value and thus required input of energy to the system The negative entropy change ΔS^* for this step of the reaction reflects the change of lectin - glycoprotein transition complex to a more ordered structure. This positive value of ΔH^* shows that the heat content of the activated complex is more than that of isolated species

Partial immobilization due to hydrophobically associated complex formed, in step 1 this happened when the isolated hydrated species (such as thyroid gland tumor lectin to glycoprotein) interact partially, so that there is a mutual penetration of their hydration layers to form the activated complex. The hydrophilic amino acid residues were previously accessible to solvent in the isolated sub units and to be buried upon complex formation hence produce an increase in the number of released water molecules.

Thus, the negative value of ΔS^* refer to the loss of a number of transitional and rotational degree of freedom originally present in both isolated species or the water molecules that have been ordered will then be released and gain freedom of motion. while the step B the hydrophobically - associated species, participates in further interactions as illustrated in the product step C, Hence giving the fully interacting associated species, for that step 2 is description of intermolecular interactions between lectin and its glycoprotein .

The formation of protein - protein or protein - ligand complex is proposed to occur in two steps , the first step stabilized by hydrophobic interactions and the second step stabilized by short - range interactions such as ionic (electrostatic) interactions, protonation , hydrogen bonding and Vander Waals interaction [204].

The hydrohpobic interaction contribution due to the stability of the complex via high positive entropy changes ($\Delta S^* > 0$) whereas the ionic interactions hydrogen bonding, protonations and Vande Waals interactions stabilize the complex by the negative change in ΔS° [204-205] and the evidence from this investigation besides the other investigation studies [205-206], indicate that this model may be useful for understanding the interaction of lectin with glycoproteins and with small molecules.

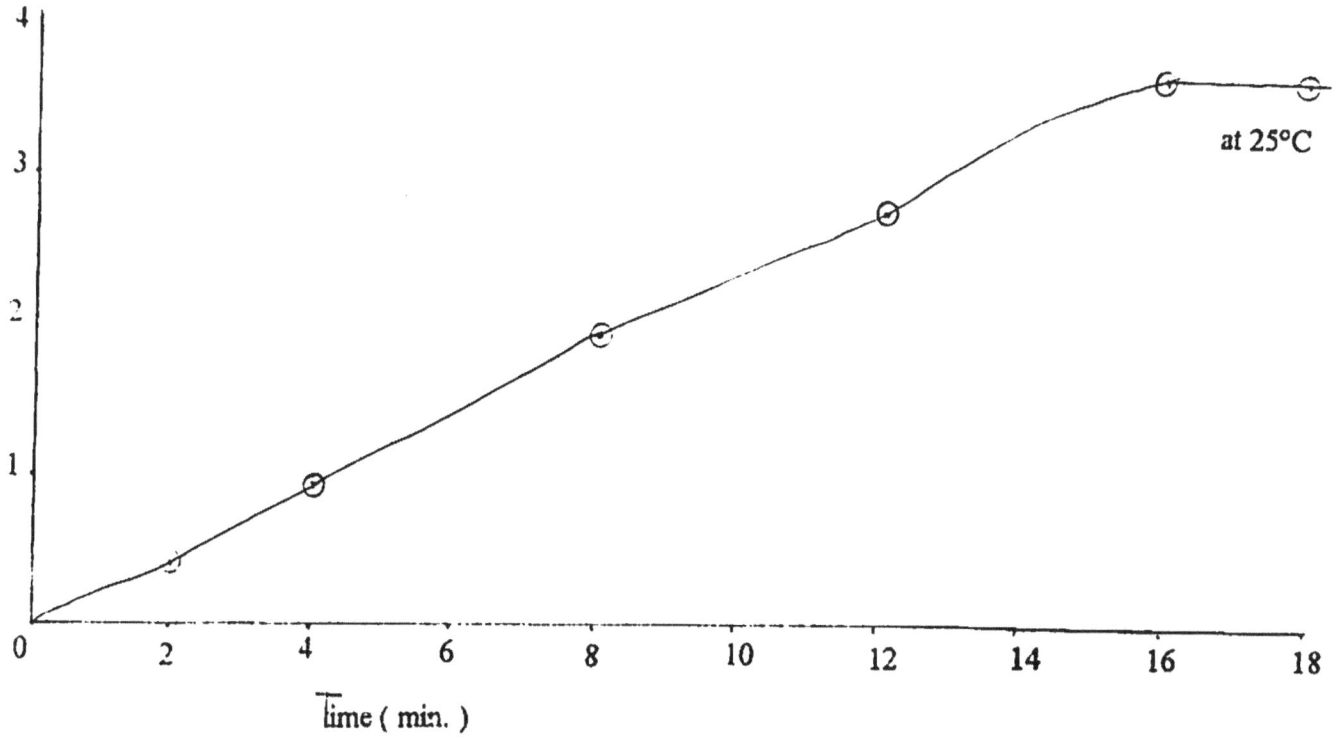

Fig. (3-20a) : The time course for association of thyroid gland tumor lectin to glycoprotein . All details are explained in text.

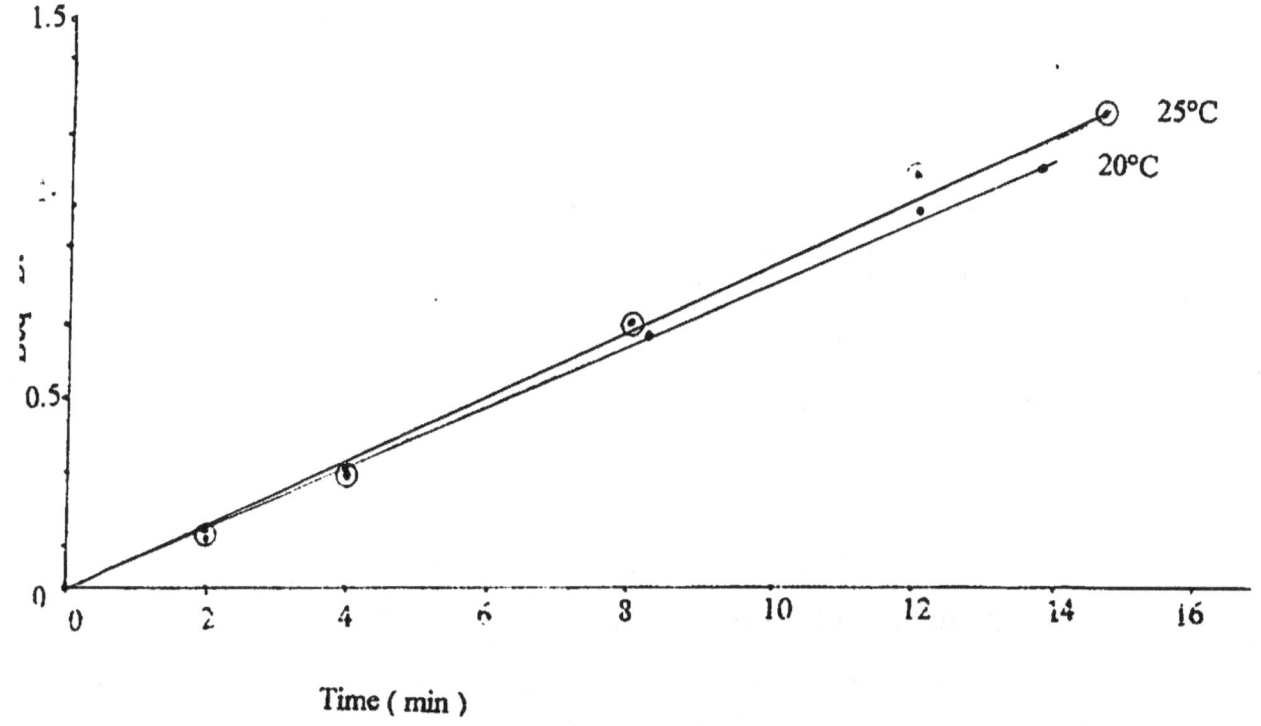

Fig. (3-20b) : The kinetic of complex formation between lectin and glycoprotein at 25°C , pesudo first order kinetic of binding .

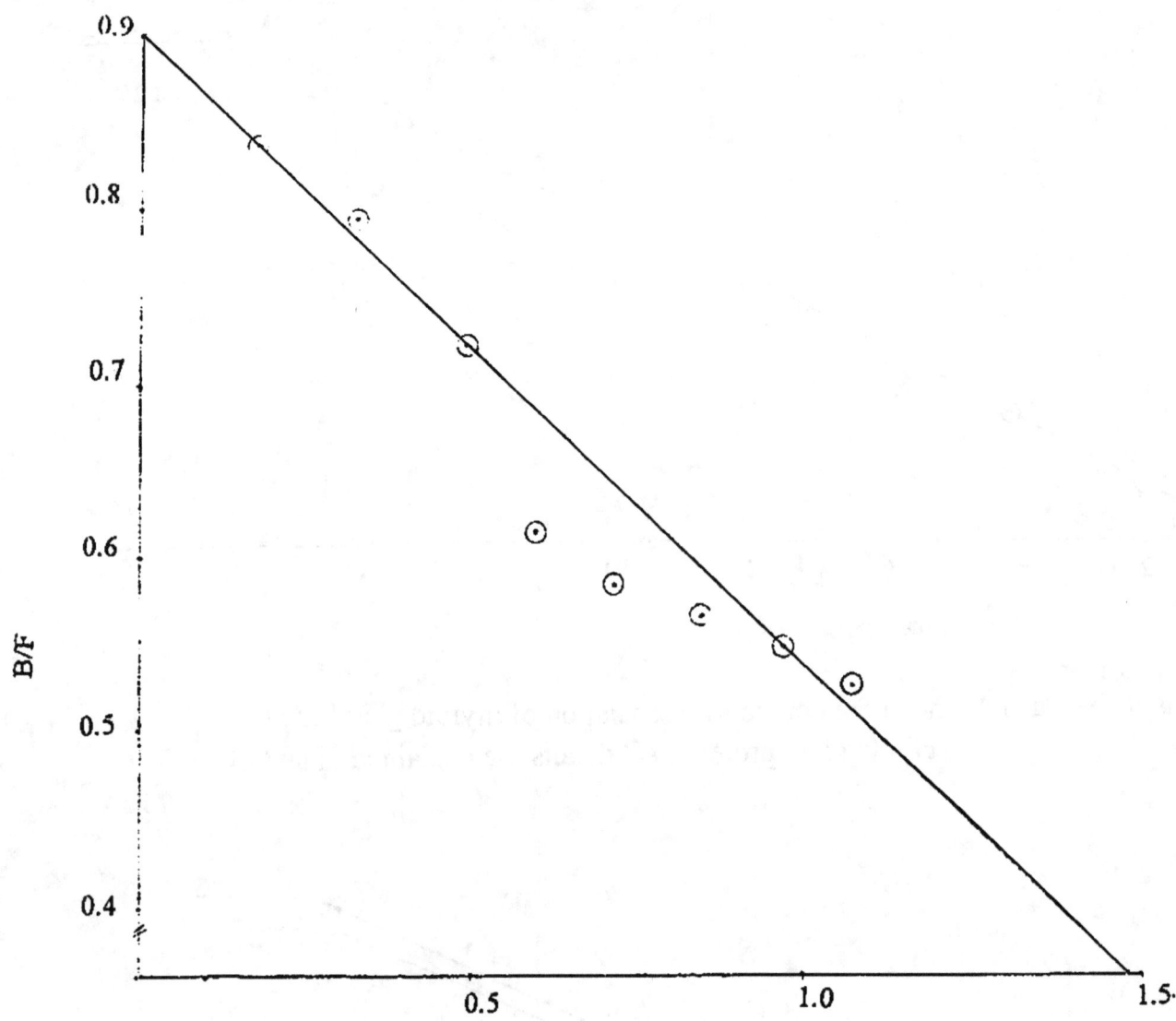

conc. Of B (M)X 10-7.

Fig. (3-21) : Scatchard plot of specific binding thyroid gland tumor
lectin to red cells glycoprotein.

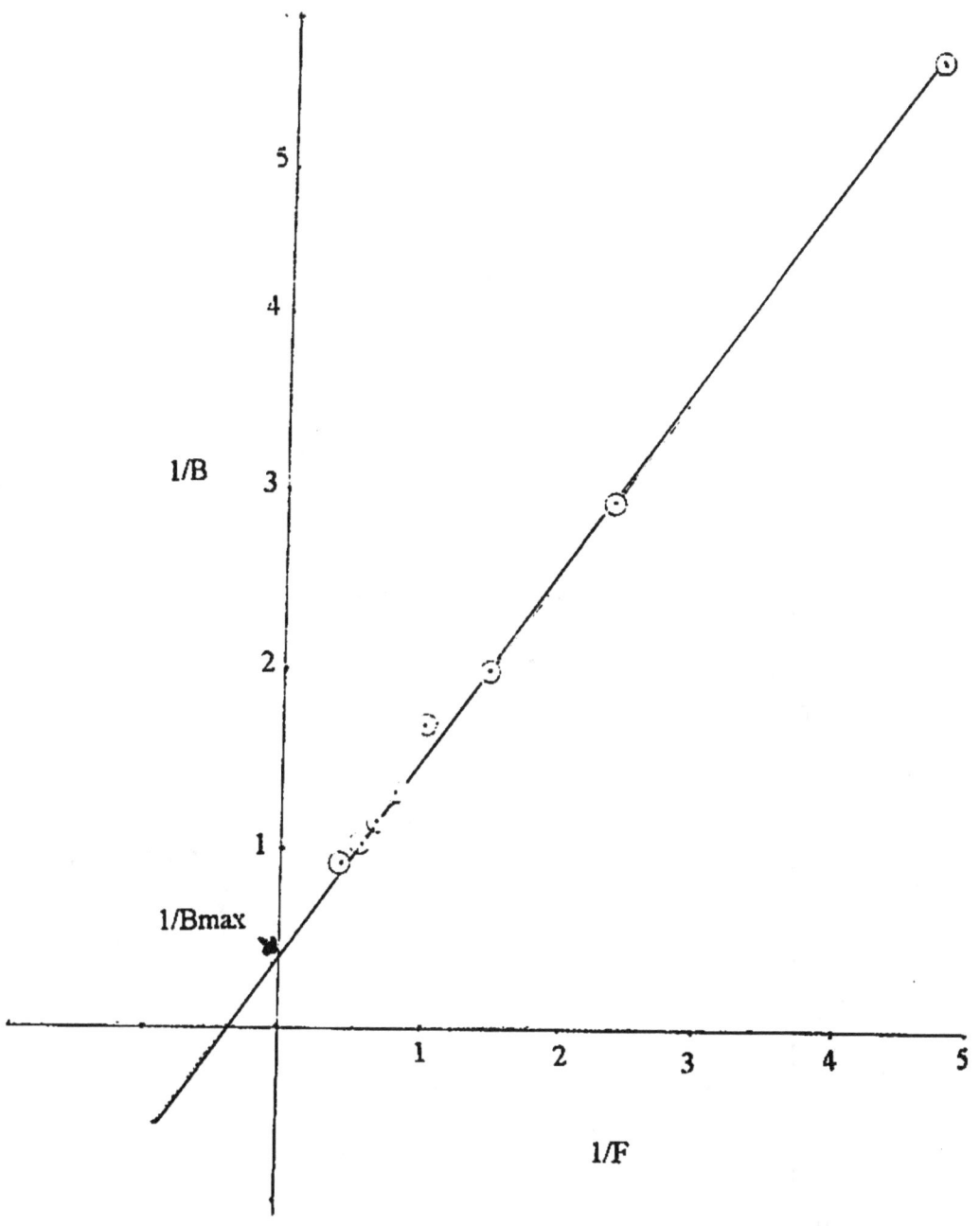

Fig. (3.22) : Line weaver - Bruk plot of specific binding of thyroid gland
tumor lectin to red cells glycoprotein .

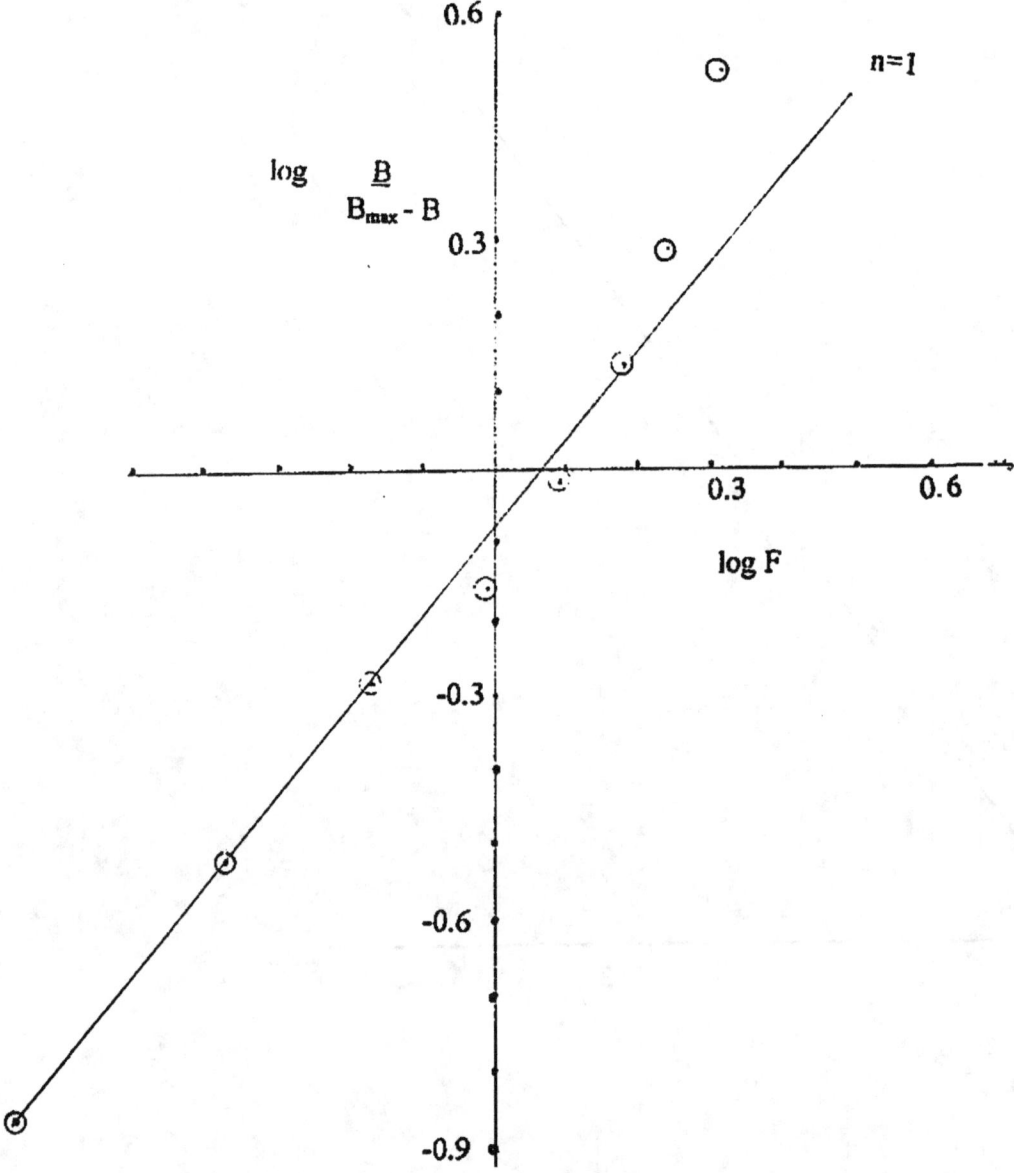

Fig. (3-23) : Hill plot of the binding of thyroid gland lectin . All details

in text.

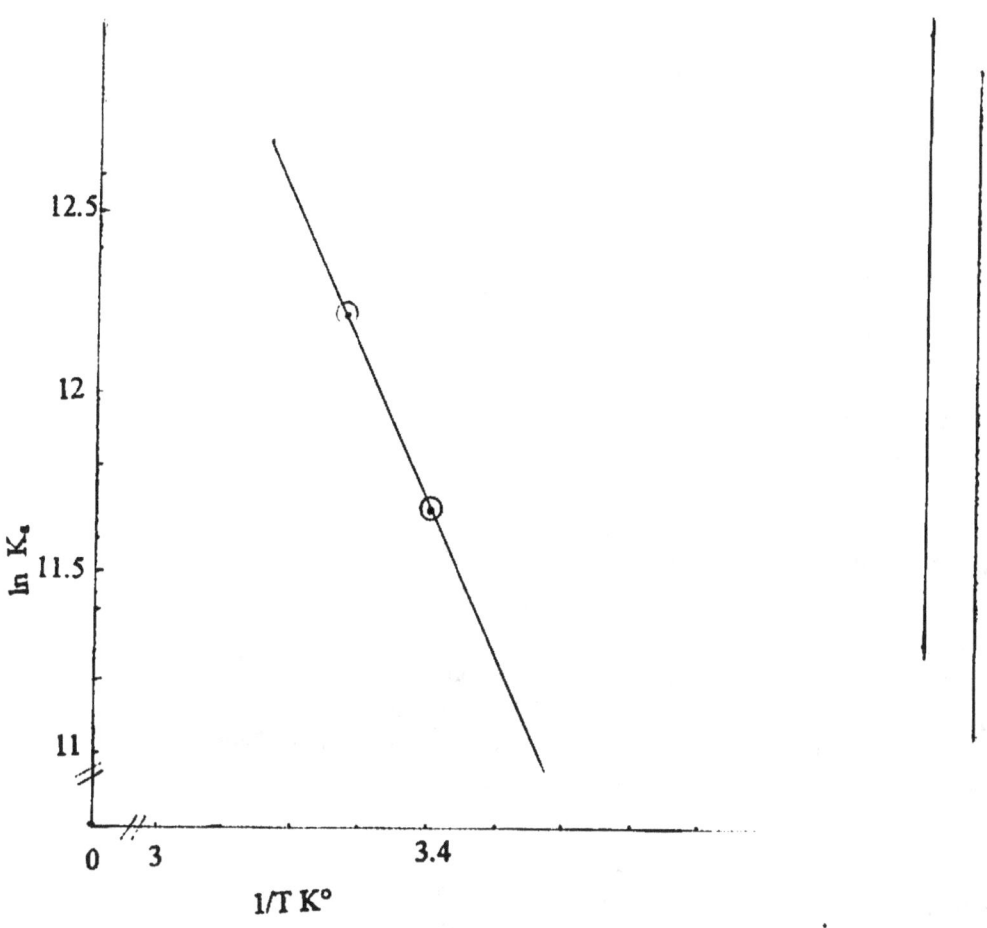

Fig. (3-24) : Arrhenius plot for the binding of thyroid gland tumor lectin
to glycoprotein .

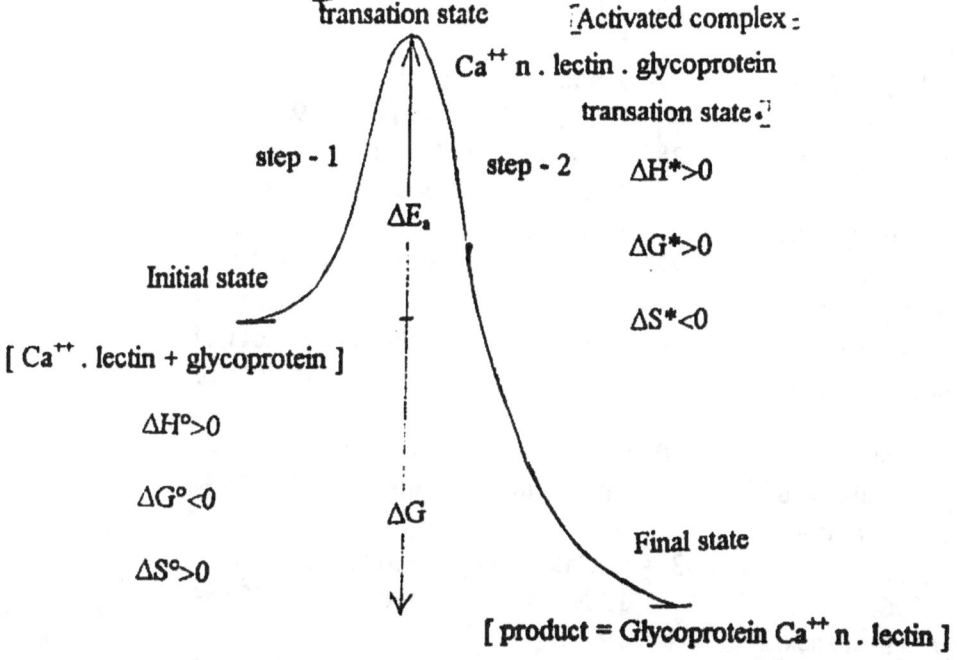

Fig. (3-25) : General energy diagram and thermodynamic model applied
to the interaction of the lectin with glycoprotein .

REFERENCES

1- Minton, J.P. and Chevinsky, A. 1989. Serum Surg . oncology . 1.5: 426.

2- Kaplan. 1989. Clinical biochemistry Theory analysis . 2nd Ed .

3- Lagara, A.; Martinez B.P. ; Marino. A.; Fago, G. and Bizzari . M.1995. Anti Cancer Res . Sep. - Oct . 15 (2341 - 6).

4- Lapez, Saes, J.J. ; Senra , Varela, A. 1995 . Int. J.Biol. markers, Jul. - Sep. ; 10(3) pp (174-9) .

5- Dai, J; Allard. WJ; Davis, G ; Yeung, KK. 1998 Tumour Biol 19 Suppl 1:100-10

6- Ivan Roitt . 1988 . Essential immunology . Sixth Ed. Black well scientific publ. 101.

7- Reid, P..E.; Culling, C.F. and Dunn, W.L. 1978. J . Histochem. 26:187-192 .

8- Culling, C.F. and Reid, P.E. 1979. J. Histochem. Cytochem. 27: 177-1179 .

9- Horowitz, M.I. and Pigman, W. 1977. 1st. Ed. Academic press. vol.1 189-213 .

10- Thomas, M. Devlin. 1992. Text book of biochem 3rd Ed. Weliy press p. 417.

11- Schauer , R. 1982. Adv. Carb. Bioch. 40 , 234.

12- Patel, V. 1978.The glycoconjugates. Academic press . vol. 2 . 185-229.

13- Miguel, A.; Ferrero; Angel Reglero; Manual Roberto, O. and Leandro, B. 1996. Great Brit. J. Biochem. 317:157-162.

14- Alfred Gottschalk . 1960 . The Chemistry and biology of sialic acid and related substances . Cambridge Un press p. 33 .

15- Alfred Gottschalk . 1960 . The chemistry and biology of sialic acid and related substances . Cambridge Un. press .p. 88.

16- Alfred Gottschalk. 1960. The chemistry and biology of sialic acid and related substances. Cambridge Un. press .p. 26.

17- Grossman, M.; Wong, R.; the. N.G.; Tropea. JE; East. Plamer. J. 1997. Endocrinology Ja; 138(1): 92-100 .

18- Oho. T; Yu, H; Yamashita, Y; Koga, T, 1998 Infect –Immun Jan; 66 (1) : 115-21.

19- Lawrence, M. et al. 1977. Clin. Chem. 23: 2055.

20- Horowitz, MI. and Pigman, W. 1977. 1st Ed. vol. 1 Academic press. 189-213.

21- Okada, Y. and Spino, RG. 1980. J.Biol. Chem. 1255, 8865 - 8872.

22- Edge, A. S.B. 1984. J. Biol. Chem. 259: 4710-4713.

23- Bossman, NB. 1972. Biochem. biophy. Acta . 279, 456.

24- Watkins, E. & et al. 1974. Int. J. Cancer. 14,799.

25- Gambaryan, AS & et al. 1977. Vibrology Jun. 232 (2) : 345-50 .

26- Daniels & et al. 1986. Vox Sang. 50 : 17.

27- Zetta, P. & et al. 1997. J. Inorg Bioch. Feb. 1 : 65 (2) : 109 - 14.

28- Bahl, OP. and Shah, RH . 1977. The glyco conjugates. 1st. ed. Academic press vol. 2 : 385-422.

29- Albert, S.B. and Robert, G. 1985. The Journal of biolog. Chem. by American Society of biological Chem. Inc. vol. 260 . no. 28 . p. 15332.

30- Helton, TE. and Magner, JA. 1994. Endocrinology Jun. 134(6): 2347-53.

31- Klaus Jung, Monika Pergand. 1989. Clin. Chem. 35/9, 1955-1957.

32- Hoermann, R.; Keutmann, HT. and Amir, SM. 1991 . Endocrinology. Feb : 128(2) : 1129-35 .

33- Wasserman. RL. and Capra. JD. 1977. The glycoconjugates. 1st. Ed. Academic press vol. 1 p. 323-348.

34- Greenhalt, T. & et al. 1973. Brit. J. Haematol. 25, 207.

35- Grossmann, M., Wong, R.; NG; Tropea, JE. 1997. Endocrinology Jan 138 (1): 92-100.

36- H Jame Harwood JR; Lorraine. D. 1997. Biochem J. Great Brit. 323, 649-659.

37- Hakomori, SI., 1984. Annu. Rev. Immuno. 2, 103.

38- Hulbert, KB.; Karim, A. Elizabeth, L. and Feranado. A. 1979. Cancer Res. 39: 5036-42.

39- Edge. AS.; Spiro, RG.1985. J. Biol. Chem. Dec. 5; 260(28): 15332-8.

40- Komminoth, P. ; Roth, J. and Saremaslani, P. 1994. Am. J. Sury .Path. Apr. 18(4): 399-411.

41- Horgan, I. 1982. Clin. Chem. Acta. 118, 327.

42- Silver, H. et al .1983. Int. J. Cancer. 31:39 .

43- Prof. Dr. Almuddaffar and Naser. 1995. Dep. of Chem. Baghdad UN.

44- Prof. Dr. Almuddafar and Naser .1995. Dep. of Chem. Baghdad Un.

45- Hirshaut, Y. et al. 1981. Proc. Am. Assoc. Cancer Res. 22, 186.

46- Hirshaut , Y. et al .1981. Proc. Am. Assoc. Cancer Res. 22, 186.

47- XinLi. ZHU.; Chun, LUO.; Jack, M.and Jaro Sodek.1997. Biochem. J. Great Brit. 323; 637-643.

48- Erbil, K. et al. 1985. Cancer 55,404.

49- Kim, Y. and Isaacs, B. 1975. Cancer Res. 35, 2092.

50- Albert, KB.; Silver, MD. and Donald, I. 1978. Cancer. 41:1497-1499.

51- Dnistrian, A. and Schwartz, M. 1981. Clin. Chem. 27,1737.

52- Dwivedi. C.; Dixit, M.; Hardy, RE. 1990. Dep. of Pharm South

53- Khaderia, V. and Keller, J. 1983. J. Surg . Oncol. 23: 163-166.

54- Harvey et al. 1981. Glycoproteins and human cancer. J. Cancer. vol. 47 : 324-327.

55- Nonda et al.1982. J. Cancer Res. 42. 5270.

56- Samuel, H. Barondes. 1984. J. Scince vol. 223 no. 4639-4643.

57- Raymond Gantt, Seymon miliner and Binkely.1964. J. Biochem. 1952-1960.

58- Kokoglu, E.; Sonmez, H. and Uslu, E. 1992. J. Cancer Biochem.

59- Alfred Gottschalk. 1960. The chemistry and biology of sialic acid and related substances. cambridge Un. press p.88.

60- Martin, DW. Jr. 1985. Harpper's review of biochem. 20th ed Lang. med. pub. p. 464-479.

61- Dnistrian, A. and Schwartz, M. 1981 . J. Clin. Chem. 27: 1737.

62- Franchin, G. et al . 1997. Infec. Immun. 65(7): 2548-54.

63- Laurent, B.1958. Scandinav. J. Clin. Invis. 10,1.

64- Montreuli, J.1957. Chem. rt Biochimi. Bull. Soc. Chim Med. 30,3.

65- Schultz, H. 1958. Dtsch Med Wochschr. 83, 1742.

66- Ng, RCY.; Roberts; An Wilson, RG. 1987. Br. J. Cancer. 5513: 249-254.

67- Vrethem, M.; Ohman, S.;Von Schench, H.1987. Acta Neural Scan. 7515: 328-331.

68- Garrausay, K. and Spielman, J. 1986. Mol. Cell Biochem. 72, 109.

69- Codington, J. et al .1978. J. Natl. Cancer Inst. 60, 811.

70- Codington, J. and Frim, D.1983. Biochembrans. 11,207.

71- Spiro , RG. ; Bhoyroo , VD. 1988 . J. Biol. Chem. Oct. 5 ; 263(28) : 14351-8 .

72- Lodish, H. F. 1989. J. Biol. Chem. 263: 2107-2110.

73- Hirschberg, CB. and MD. Snider. 1987. Annu. Rev. Biochem. 56: 63 -88.

74- Horowitz, MI. and Pigman. W. 1977. 1st ed.. Academic press. vol. (1) : p. 1-10 .

75- Kameda, Y. 1992 . J. Histochem. Cytochem. Apr. 40(4): 541-53.

76- Robert, K. et al. 1988. Harpper's Biochem. 21st ed. Middle east. p. 589.

77- Devlin. 1992. Text book of biochem. Third Ed. Weliy Less press.

78- Rellier, N. et al. 1997. Biochem. Biophy. Res. Com. Jun. 18: 235(2): 281-5.

79- Vliegenhart, JFG. and H van Hal beek. 1983. Adv. Carbohy. Chem. Biochem. 41: 209-374.

80- Jeffrey, A., Gary Boss, Timothy Weaver, Ward Rice, Carol Dion and William Hull. 1985. j. of Biology. Chem. vol. 260 . no. 28. p. 15273.

81- Lenten, D.V. and Ashwall, G. 1970. J. of Biol. Chem. 246: 1889-1894.

82- Schwick, H.G.; Heide, K. and Haupt, H. 1977. The glycoconjugate. vol. Academic press. 261- 321.

83- Selenkow, H.A.; Birubaum, M.D.; Hollande, CS. 1973. J. Clin. Obstest Gynecol. 16: 66-68.

84- Feizi, T. and Childs, RA. 1985. Biochem . Sci. Jan. p. 24-29.

85- Lawrence, A. Kapplin. 1989 . Clinical Chem. 2nd ed. Mosby Company p. 757.

86- Kilm, S. and Schauer, R. 1997. Int. Rev. Cytol. 175 : 137-240.

87- Macchia, E.; Concetti, R.; Carone, G.; Borgoni, F.; Fenzi, GF. and Pinchera, A. 1988. J. Article Feb. 28(2); 28(2): 147-56.

88- Iwatani, Y.; Litaka, M. et al. 1987. J. Clin. Endocrinol . Metab . Jun; 64(6): 1302-8.

89- Farmaniak, J., Daveenport, S. et al. 1988. Clin. Endocrinol. Jun; 28(6): 589-600.

90- Udo Schumacher; Hans, P.H.; Ulrich and Edwin, K. 1989. Histochemical J. 21: 44-46.

91- Judd, W.J. 1980 . Crit. Rev. Clin. Lab. Sci. 1 : 171-214.

92- Finne, J. 1980. J. Biochem . 104: 181-189.

93- John bernard, MD. 1984. Clinical diagnosis and managment. 17th Ed. Sanders press. p. 981.

94- Chitra, I. and Sujata, B.1987. Biochem. and Biophys. Res. Commun. vol. 148 . no. 2:795-801.

95- Fredrich. 1996. Biochem. j. 316: 123-129.

96- L. Bladier, D. and Aminoff , D. 1986. Natl. Acad. Sc. USA 83, 5:1339-1343.

97- Ronald L.; Miller; james, F.; Collawn, Jr. and W. Fish.1982. J. Biological Chem. vol. 257. no. 13 II0: 7574-7580.

98- Nicolson, Gl. 1976. Biochem. Biophys. Acta 457: 57-108.

99- Sasano, H.; rojas, M.; Silverberg, SG. 1989. J. Article Feb. 113(2): 186-9.

100- Bhavandan and Katlic . 1979 . J. Biol. Chem. 254 : 4000-4008 .

101- Samul, H. 1984. Science. 1223-1259.

102- Mathews. 1990. Biochemistry, the Benjamin cummings Pub. p. 294.

103- Gonzalez Campora. 1988. Cancer Dec. 1, 62, (11) : 2345-62.

104- Souj - EDA, Yasuhiko Suzuki. 1996. J. Biochem. Print in Brit. 3169: 43-48.

105- R.C. Hughes. 1983. Outline studies in biology of glycoprotein. p.69.

106- Robert, A. et al. 1984. Cancer Res. 22: 170.

107- Schauer, R. 1982. Advanc. Carbohyd. Chem. J. Biochem. 40: 131-234.

108- Schauer, R. et al. 1980. Structure and fuctions of gangliosides, Planum press. Nt. p. 283-294.

109- Varki et al. 1980. J. Exp. Med. 152 : 532-544.

110- Chitra and Sujata Basu. 1987. Biochem and Biophys. Res. Comm. v. 148 no. 2.

111- Barodes, S. 1981. J Annu. Rev. Biochem. 80: 207 .

112- Fredrick, et al. 1996. Biochem. J. 316: 123-129.

113- Goldstein, R. 1980. J Nature. 285-66.

114- Bouchara, JP. et al . 1997. Infect. immun Jul. 65(7): 2717-24.

115- Krach et al. 1974. Exp. Cell Res. 84-91.

116- Souji, EDA and Yasuhiko Suzuki. 1996. Biochem. J. 3169: 43-48.

117- Lin, J. Y.; Li, J-S. and Tung, T. 1981. Biochem. Biophys. Acta. 638: 275-281.

118- Lawrence, A., Kaplan. 1989. Clinical Chem. 2nd ed. Mesby comp. p. 181.

119- Mridula, C., Manju, S. and Chitra. M. 1985. Biochem. and Biophy. Res. Com. vol. 130, 3.

120- Kharana, S.; Raghunathan, V. 1997. Biochem Biophy. Res. Com. May 19, 234 (2): 465-9.

121- Nowak, TP. and Barondes, SH .1975. Biochem. Biophys. Acta. 393: 115-123.

122- Pardoe, GI et al. 1976. J. immunology. 18: 73-83.

123- Kominoth, P.; Roth, J.; Saremaslani, P.; Matias - X; Wolfe - Hjand Heitz - PU. 1994. J. Surg Path Apr. 18(4): 399-411.

124- Chitra, J. and Sujata Basu. 1987. Biochem. and Biophys. Res. Com. 148 (2) 795-801.

125- Ronald, L.; Miller James, F.; Collawn. Jr. and Wayne, W. 1982. J. Biological Chem. 257(13): 7574-7580.

126- Wiener, A.S.; Socha, W. 1974. Immunol. 47: 547.

127- Wasserman , R.L. and Capra, J.D. 1977. The glycoconjugates. 323-348.

128- Baumann, H. and Doyle, D. 1984. Molecular and chemical characterization of membrane receptors . 125-160.

129- Stern, P.L.; Willison, K.R.; Lennox, E.; Galfre, G.; Milstein, C.; Secher, D.; Ziegler, A. and Springer, T. 1978. J. Cell. 14:775-788.

130- Brollmar, E.F.; Sagi, M.; Shimura, Y.; Lau, J.T. and Ashwell, G. 1993. J. Biol. Chem. Feb. 268(5): 3604-9.

131- Stephen, s.; Sternberg. 1989. Diagnostic Surgical pathology VI, Raven press New York 395.

132- Thomas, R. 1984. Head and neck imaging 519.

133- Lawrance, Kaplin.1989. Clinical Chemistry, Theory analysis 2nd. Ed. p. 621.

134- Montani, V; Taniguchi, SI & et al 1998 Endocrinology Jan 139 (1): 280-9.

135- Balkan, W; Tavianini, MA; Gkonos, PJ; Roos, BA. 1998. Endocrinology Jan:139 (1):252-9.

136- Chopra, I.J. 1981. Triiodothyronines in health and diseases, 1st Ed. p 9.

137- Peak, Cates and Deiss. W.P. 1970 . J. Endocrinology. 87: 494 - 497.

138- Siperstien, A.E., Miller, R.A. 1988. J. Surgery Dec. 104(6): 985-91.

139- Gross, J.L. ; Vasques , I. 1996 . J. Endocrinology Invest Jan . 19(1) : 21-4 .

140- Sairam, M.R. and L. SH. 1973. Biochem and Biophys. Res. Com. 51: 336-342.

141- Brollman, Ef.; Sagi, M.; Shimura, Y.; Lan ,J.T.; Ashwell, G. 1993. J. Biol. Chem. Feb. 268 (5): 3604-9.

142- Copper, J.R., et al. 1978. Biochemical basis of neuropharmacology Oxford Un. press Inc. p. 60.

143- Hechter, O. 1978. The receptor concept Int. Plenum press. New York V. 96 p.1.

144- King, R.J.B. 1987. J. Endocrinology. 114, 341.

145- Walter, M.R. 1985. J EndocrinologyRev. 6 : 5121.

146- Chopra, I.J.; Nelson, J.C.; Solomon, D.H. and Beal, G.N. 1971. Production of antibodies specifically binding triiodothyronine and thyroxine. J. Clin. Endocrinology Metab. 32: 299-308.

147- Reftoff, S. 1979. Thyroid function test Endocrinology. 1: 387-401.

148- Ennis, B.W. 1986. Endocrinology. 119: 2066.

149- Prof. Al-Muddaffar and Alaa. 1989. Bag. Un. Dep. of Chem.

150- Hall, R. 1970. Hyperthyroidism pathogenesis and diagnosis. Brit. Med. J. 1: 743-747.

151- Prof. Al-Muddaffar and Majed . 1996. Bag. Un. Dep. of Chem.

152- Harper and Raw. 1978. The thyroid fundamental and Clinical Text 4th. Ed. 389 - 393.

153- Lawrence , A. , Kaplan , A. 1989. Clinical chemistry Mosby comp. p.632.

154- Lowery, O. et al. 1951. J.Biol. Chem. 193, 265.

155- Katapoids, N. et al. 1982. Cancer Res. 42, 5270.

156- Richard, J. 1965. Clinical Chem. principle and Tech.

157- Hans, J.; et al. 1989. Bioche. Biophys. Res. Commun. 163, 506.

158- Lineke, I. 1955. arch Biochem . Biophys. 54. 223.

159- Lis-Hand & Sharon, N. 1972. Methods on Enzymology. vol. 28, p. 360.

160- Csizman, L. 1960. Proc. Soc. Exp. Biol. - N.Y. 103, 157.

161- Nowak, T. and Barondes , S. 1975. Biochemica et Biophys. Acta. 393, 15.

162- Laemmli, U. 1970. Nature. 227, 680.

163- Pharmacies Fine Chemical; Gel filteration Calibration Kit Instruction Mannual for protein molecular weight determination .

164- Gross, J.L; Vasques. 1996. J. Endocrinology. Invest Jan. 19(1): 21-4.

165- Gol'dburt, N.N.; Perohuk , B.D.; Markin, S.S. ; Makeedonskaia I.V. 1996 Arkh, Pathology. 58(1): 56-8.

166- Bo Petterson; Michel-P-Colman; Elainc Ron and Hans Adami. 1996. Int. J. Cancer. 65: 13-19.

167- Morchek, J.; et al. 1976. Clin. Chem. 22, 1516.

168- Shamberger. 1984. J. Clin. Chem. Biochem. 22 , 647.

169- Silver, H. 1980 . Br. J. Cancer. 41, 745.

170- Grook, J. R. 1997. Clin. Cardiol. May. 20: 455-8.

171- Ann, M. et al. 1982. Cancer. 50, 1815.

172- Kim, Y. and Isaacs, B. 1975. J. Cancer Res. 35, 2092.

173- Stefenelli, N. et al. 1985. J. Cancer Res. Clin. oncol. 109, 55.

174- Collm, J.; Serratosa, J.; Baches, O. 1987. Clinical Biochem. vol. 74 I 8, p. 472.

175- Cooke, R.P.D.; Wells, F.E. and Rose, P.E. 1986. Med. Lab. Sci. 1986 . 43/14 (382-385).

176- Reid, P. et al. 1975. Can. J. Biochem. 53, 1328.

177- Yamamooto, K. et al . 1984. Eur. J. Bioch. 143, 133.

178- Catchpole, H. 1950. Proc. Soc. Exp. Biol. Med. 75, 221.

179- Yaskhiko, T. et al. 1988. Am. J. Nephrol. 2, 21.

180- Sheltar, M. et al. 1949. Proc. Soc. Exp. Biol. Med. 72, 294.

181- Bohmer, S. and Davidson, E. 1981. Biochemistry. 20, 1047.

182- Seibert, F. et al. 1947. J. Am. investig. 26, 90.

183- Maria, G. et al. 1989. Biochem. J. 258, 569.

184- Dumont, J.E. and Vassart,G. 1979. Endocrinology, Grune and Staratton New York.:311-329.

185- Tate, R.L.; Hollmes. J.M.; Kohn, L.D. and Winand, R.J. 1975. J. Biol. Chem. 250, 6527 - 6533.

186- Meldolesi, M.F.; Fishman, P.H.; Aloj, S.M.; Ledley, F.D.; Lee, G.; Bradley, R.M., Brady, R.O. and Kohn. L.D. 1977. Biochem. Biophys. Res. Commun. 75, 581-588.

187- Mullin, B.R.; Pacuszka, T.; Lee, G. 1978. Scince. 199: 77-79.

188- Mc Neil, G.; Virji, N. 1997. Microb-pathog. 22(5): 295-304.

189- Albert, SB. and Robert, G. 1985. J. Biol. Chem. vol. 260. no. 28. IS pp. 15332-15338.

190- Hans , J. et al. 1989 . Biochem . Biophys . Res. Commun. 163, 506.

191- Gambino, R. et al. 1997. J. Protein Chem. Apr. 16(3): 205-12.

192- Mathews. 1990. Biochemistry Benjamin press, p. 293 .

193- Isamu Matsumoto; et al. 1986. Carbohy Res. 151, 261-270.

194- Elvin, A. and Manfred, M. 1967. Experimental Immuno. Chem. Second Ed. T Linois, USA.

195- Prof. Al-Muddaffar, S. and bakhtear. 1992. Coll. of Science, Baghdad Un.

196- Mancini, G.M.S.; et al. 1989. J. Chem. 264, 26: 15247-54.

197- Goebel, W.; et al. 1934. J. Exper. Med. 60, 599.

198- Wild. J.; et al. 1983. Biochem. J. 21, 167.

199- Dolichos, R. 1983. Biochem. 22. 2741.

200- Georg Lee; Evelyn, F.; Grollman; Sherry Dyer & et al. 1970. Biolog. Chem. vol. 245. no. 10 I 25 p. 3826-32.

201- Gambino, R.: et al 1997. J. Protein Chem. Apr. 16(3): 205-12.

202- Tark, H. 1992. Ph.D. thesis, College of Science, Baghdad Un.

203- Rae-Venter, B. and Dao, T. 1982. Biochem. Biophys. Res. Comm. 107, 624.

204- Hussain, M. 1990. Msc. thesis, College of Science Un. of Baghdad. pp.87.

205- Blumenthal, D. and Stull, J. 1982. Biochemistry. 21, 2386.

206- Laporte, D. et al. 1980. Biochemistry. 19,3814.

207- Tobo, T; Oktomo, N; Vinores, MA;Derevjanik, NI; Vinores, SA; Zak, DG; 1998, Invest. Ophthalmol. Vis. Sci. Jan; 39(1): 180-8.

208- Montani,V; Shong, M; Taniguchi, SI; Suzuki, K; 1998:Endocrinology Jan 139 (1):290-302.